A Textbook of

Engineering Physics Practical

A Textbook of
Engineering Physics Practical

By

Dr. Ruby Das
Department of Engineering Physics
Bhilai Institute of Technology, Durg
Chhattisgarh

C. S. Robinson
Department of Engineering Physics
Bhilai Institute of Technology, Durg
Chhattisgarh

Dr. Rajesh Kumar
Department of Engineering Physics
Bhilai Institute of Technology, Durg
Chhattisgarh

Prashant Kumar Sahu
Department of Engineering Physics
Bhilai Institute of Technology, Durg
Chhattisgarh

UNIVERSITY SCIENCE PRESS
(An Imprint of Laxmi Publications Pvt. Ltd.)
An ISO 9001:2008 Company
BENGALURU • CHENNAI • COCHIN • GUWAHATI • HYDERABAD
JALANDHAR • KOLKATA • LUCKNOW • MUMBAI • RANCHI • NEW DELHI
BOSTON (USA) • ACCRA (GHANA) • NAIROBI (KENYA)

A TEXTBOOK OF ENGINEERING PHYSICS PRACTICAL

Printed and bound in India
Typeset at Abro Enterprises, Delhi
First Edition : 2010, Second Edition : 2016
ISBN 978-93-80386-86-7

Branches

Bengaluru	080-26 75 69 30	
Chennai	044-24 34 47 26,	24 35 95 07
Cochin	0484-237 70 04,	405 13 03
Guwahati	0361-254 36 69,	251 38 81
Hyderabad	040-27 55 53 83,	27 55 53 93
Jalandhar	0181-222 12 72	
Kolkata	033-22 27 43 84	
Lucknow	0522-220 99 16	
Mumbai	022-24 91 54 15,	24 92 78 69
Ranchi	0651-220 44 64	

PUBLISHED IN INDIA BY

UNIVERSITY SCIENCE PRESS
(An Imprint of Laxmi Publications Pvt. Ltd.)

An ISO 9001:2008 Company
113, GOLDEN HOUSE, DARYAGANJ,
NEW DELHI - 110002, INDIA
Telephone : 91-11-4353 2500, 4353 2501
Fax : 91-11-2325 2572, 4353 2528
www.laxmipublications.com info@laxmipublications.com

C—10586/015/07
Printed at: Repro Knowledgecast Limited, Thane

Contents

Preface to the Second Edition

Engineering, the science of measurement, is the offspring of physics which plays the primary role in each branch of engineering and is generally taught during the first two or three years of these courses in various technical universities, NITs and IITs. The facts predicted by theories are brought nearer to the truth through practicals. It is in the laboratory that physics students learn to practice the activities of scientists—asking questions, performing procedures, collecting and analyzing data, answering questions, and thinking of new questions to explore.

This book grew out of the lectures delivered in theory and practical classes by us to the students at the Bhilai Institute of Technology, Durg (Chhattisgarh). Along with the CSV Technical University practical syllabus, the book attempts to give a comprehensive account of a large number of other important experiments in engineering physics as well. The first edition of the book published in 2010 and it is indeed a matter of great satisfaction that the book has been received well by the academic community in central India. While using the book in the undergraduate laboratory course for engineering students, we felt the necessity of revising its certain portions.

In the Second Edition the following revisions have been done:

1. Several diagrams have been replaced in various chapters by blending careful presentation of fundamental concepts and methods of physics.
2. The concept of measuring least count in screw gauge is now more clear.
3. Measure of capillary tube diameter is redrawn. Corrections have been made in current-intensity characteristic graph of photoresistor and NOT gate electrical switching circuit.
4. In experiment 16 many points have been added in the procedure to make it clearer.
5. An alternative method to determine the forbidden energy gap of P-N junction diode material has been added.

The organization of the book remains the same. Any further suggestions for improvement of the its contents are most welcome.

—AUTHORS

Preface to the First Edition

The basic fundamental Physics forms the foundation of engineering and technology. Physics, and natural science (being a description of the nature around us) in general, is a reasonable enterprise based on valid experimental evidence, criticism, and rational discussion. It provides us with knowledge of the physical world, and it is an experiment that provides the evidence that grounds this knowledge. Experiment plays many roles in science. One of its important roles is to test theories and to provide the basis for scientific knowledge. Experiment can provide hints toward the structure or mathematical form of a theory and it can provide evidence for the existence of the entities involved in our theories.

This book has been written with the specific aim of presenting the subject-matter in the simplest possible manner keeping in view the difficulties and the limitations of the average students. The completely redesigned, pedagogically consistent artwork and diagrams integrate seamlessly with the text to help "more visual" learners better visualize key concepts. The difficult experiments are explained in a simple language with the help of neat diagrams. We feel that if a student does all these thoroughly, the subject will appear to be friendly and enjoyable.

It is earnestly hoped that students and teachers of physics, engineering and technology will find the book useful. The authors will appreciate any genuine and authentic suggestions, for the substantial improvement of the book. All available standard books on the subject have been freely consulted during the preparation of this book. We acknowledge gratefully our indebtedness to the authors and publishers of different books on the subject. Authors are thankful to the management of Bhilai Institute of Technology for providing all facilities in the laboratory. We thank the members of our family for their support and encouragement. Finally we are grateful to Laxmi Publications, (P) Ltd, New Delhi, for taking keen interest in the publication of this book.

—AUTHORS

Foreword

"Physics is the Father of all branches of Engineering and Technology." The physicist discovers scientific principles and invents devices to describe and explain them. The technician applies and magnifies these devices for human convenience and comfort. There is no branch of engineering or technology in which one or other physical principle is not applied on a scale never contemplated by the discoverer. Knowledge of Physics is so interwoven with engineering studies that one cannot think of pursuing engineering studies without the knowledge of Physics. Study of Physics is essential for students and practitioners of engineering and technology to develop them in proper understanding of physical phenomenon, scientific temper and engineering aptitude. Thus, it becomes immensely important to develop zeal among the students by creating an innovative ideas and thoughts. Obviously the laboratories are the best corners for inculcating such habits.

The authors have judiciously opted to jot down their ideas and work in the form of a textbook on Practical Physics which is of utmost importance to students. Such type of book was under very much demand and I am of strong belief that the demand can be fulfilled by the present book. Moreover, this book will definitely reduce the gap between the theoretical aspects of Physics and its practical counterpart which is the need of the day. In short, the book is an indispensable reference to faculty members and laboratory instructors and an invaluable aid to students.

I am pleased and honoured to have been asked to write the foreword for this book. The authors should be congratulated for providing this valuable reference book for the would be technocrats.

Dr. M. K. Kowar
Director,
Bhilai Institute of Technology,
Durg, Chhattisgarh

Assessment of Practical Skills in Engineering Physics

For BE-Ist year Examination of Chhattisgarh Swami Vivekanand Technical University

EVALUATION SCHEME

Examination	*Duration*	*Marks*
Internal (college based) Evaluation 1. Testing of practical skill 2. Record of practical results 3. Viva	1½ hrs (per week)	20
External Examination (Examiners are appointed by the University)	1½ hrs (one experiment will have to be performed)	40
Total	3 hrs	60

❑❑❑

ENGINEERING PHYSICS LABORATORY

DO'S

1. Obtain copy of the syllabus of your Engineering Physics lab.
2. Always be regular in practical classes and complete your work within the given schedule. In all twelve turns (two periods, total of 90 minutes per turn = 24 periods) shall be scheduled in the whole semester and there are ten experiments to be performed from different branches of Engineering Physics.
3. Write down the experiments in your practical file in the same manner as given in this book.
4. Handle all the instruments carefully. Students found involved in any kind of misplacement or breakage of lab instruments may be liable to fines.
5. Bring calculator, graph paper (if required) and the observation copy regularly in your lab periods. At the end of the period submission of observation copy is a must for getting attendance.
6. Concentrate on each line written in this book, go through the experimental set-ups and fill all the observation tables mentioned. Show calculations in the given manner. Prepare answers of viva-voce.
7. Display a disciplined behavior. This should reflect in your manners and even in your talks. For any quest regarding the Engineering Physics practical, discuss with Teacher in-charges.
8. The award of sessional marks in the examinations is largely dependent on the attendance in the theory and practical classes as well as in internal marks obtained by your performance in the lab, which is entered by the teachers in the enclosed internal invaluation chart for teacher incharges. So, batch-wise get your readings checked by the teacher before leaving the lab.

DON'TS

1. Do not enter the lab without your observation copy in which your name, batch number and turn number is mentioned.
2. Do not try to manipulate your readings for getting errorless results. Instead of that learn how to perform the experiments and discuss your results with the Teacher incharges.
3. Do not avoid the mentioned precautions to be measured in performing each experiment.
4. Do not request for your choice in performing experiment at the time of practical examination. Anyone experiment shall be asked.

❑❑❑

RECORD OF THE EXPERIMENTS PERFORMED

Name of the Student : ....

Class/Section : ....

Roll No. : ....

S.No.	*Experiment No.*	*Experiment Name*	*Date*	*Page*	*Signature/Remarks*

❑❑❑

KNOWLEDGE OF BASIC DEFINITIONS

METER

Meter is the unit of length in S.I. System.

Meter is defined as *"The distance between the two marks on a Platinum-Iridium bar kept at 0°C in the International Bureau of Weight and Measures in Paris."*

One meter = 100 cm

One meter = 1000 mm

KILOGRAM

Kilogram is the unit of mass in S.I. System.

"Kilogram is defined as the mass of a platinum cylinder placed in the International Bureau of Weight and Measures in Paris."

One kilogram = 1000 gram

SECOND

Second is the unit of time in S.I. System.

A second is defined in terms of the time period of Cs-133 atoms.

i.e., "one second is equal to 9,192,631,770 periods of vibrations of Cs-133 atoms."

60 seconds = one minute

3600 seconds = one hour

AMPERE

It is the unit of current. One ampere is that constant current which, if maintained in two straight parallel conductors of infinite length, of negligible circular cross-section, and placed one meter apart in vacuum, would produce between these conductors a force equal to 2×10^{-7} Newton per meter of length.

KELVIN

The fraction 1/273.16 of the thermodynamic temperature of triple point of water is called 1 K.

OHM

The ohm is the electric resistance between two points of a conductor when a constant difference of potential of 1 volt, applied between these two points, produces in this conductor a current of 1 ampere, this conductor not being the source of any electromotive force.

COULOMB

The Coulomb is the quantity of electricity transported in 1 second by a current of 1 ampere.

MOLE

The amount of a substance that contains as many elementary entities (molecules or atoms if the substance is mono-atomic) as there are number of atoms in 0.012 kg of carbon-12 is called a mole. This number (number of atoms in 0.012 kg of carbon-12) is called Avogadro constant and its best value available is 6.022045×10^{23} with an uncertainty of about 0.000031×10^{23}.

CANDELA

The S.I. unit of luminous intensity is 1 cd which is the luminous intensity of a blackbody of surface area 1/600,000 m^2 placed at the temperature of freezing platinum and at a pressure of 101,325 N/m^2, in the direction perpendicular to its surface.

NEWTON

The Newton is that force that gives to a mass of 1 kilogram an acceleration of 1 meter per second.

JOULE

The joule is the work done when the point of application of a constant force of 1 Newton is displaced a distance of 1 meter in the direction of the force.

WATT

The watt is the power that gives rise to the production of energy at the rate of 1 joule per second.

VOLT

The volt is the difference of electric potential between two points of a conducting wire carrying a constant current of 1 ampere, when the power dissipated between these points is equal to 1 watt.

WEBER

The Weber is the magnetic flux that, linking a circuit of one turn, produces in it an electromotive force of 1 volt as it is reduced to zero at a uniform rate in 1 second.

LUMEN

The lumen is the luminous flux emitted in a solid angle of 1 steradian by a uniform point source having an intensity of 1 candela.

FARAD

The farad is the capacitance of a capacitor between the plates of which there appears a difference of potential of 1 volt when it is charged by a quantity of electricity equal to 1 coulomb.

HENRY

The Henry is the inductance of a closed circuit in which an electromotive force of 1 volt is produced when the electric current in the circuit varies uniformly at a rate of 1 ampere per second.

RADIAN

The radian is the plane angle between two radii of a circle that cut-off on the circumference an arc equal in length to the radius.

STERADIAN

The steradian is the solid angle that, having its vertex in the center of a sphere , cuts-off an area of the surface of the sphere equal to that of a square with sides of length equal to the radius of the sphere.

LEAST COUNT

Minimum measurement that can be made by a measuring device is known as "LEAST COUNT".

Smaller is the magnitude of least count of a measuring instrument, more precise the measuring instrument is. A measuring instrument cannot measure anything whose dimensions are less than the magnitude of least count.

BACKLASH ERROR

Slackness in the tilting of screw and nut.

If the direction of the movement of the screw is changed in opposite direction, the head moves through an appreciable angle before the screw begins to move. This error is called as backlash error.

ZERO ERROR

It is a defect in a measuring device (Vernier Callipers and Screw Gauge).

When jaws of a vernier callipers or screw gauge are closed, zero of main scale must coincide with the zero of vernier scale or circular scale in case of screw gauge.

If they do not coincide then it is said that a zero error is present in the instrument.

TYPES OF ZERO ERROR

Zero error may be positive or negative.

A positive zero error in the instrument shows a larger measurement than the actual measurement. In order to get exact measurement, positive zero error is subtracted from the total reading.

A negative zero error in the instrument shows a smaller measurement than the actual measurement. In order to get exact measurement, negative zero error is added to the total reading.

PITCH

"Perpendicular distance between two consecutive threads of the screw gauge or spherometer is called PITCH."

ERROR

An error is defined as *"The difference between the measured value and actual value."*

If two persons use the same instrument for finding the same measurement, it is not essential that they may get the same results. There may arise a difference between their measurements. This difference is referred to as an "ERROR".

TYPES OF ERROR

Errors can be divided into three categories :

(*i*) Personal Error (*ii*) Systematic Error (*iii*) Random Error.

Personal Error

An error comes into play because of faulty procedure adopted by the observer is called "Personal Error". Personal error comes into existence due to making an error in reading a scale. It is due to faulty procedure adopted by the person making measurement.

Systematic Error

The type of error arises due to defect in the measuring device is known as "Systematic Error". Generally, it is called "ZERO ERROR". It may be positive or negative error. Systematic error can be removed by correcting measurement device.

Random Error

The error produced due to sudden change in experimental conditions is called "Random Error".

For example, Sudden change in temperature, change in humidity, fluctuation in potential difference (voltage). It is an accidental error and is beyond the control of the person making measurement.

❑❑❑

Knowledge of Basic Measuring Instruments in Lab

Let us consider the measurement of length. We use a meter scale for this purpose. The smallest division on the scale is 1 mm. Therefore, we can measure a length to an accuracy of 1 mm with the meter scale. We can at the best estimate up to 0.5 mm. But to measure a length, we have to take two readings and their differences. Therefore, the accuracy of the scale cannot be better than 1 mm.

Fig. 1

For achieving greater accuracy, a vernier is employed. The vernier consists of an auxiliary scale called vernier scale which slides along an ordinary scale called the main scale as illustrated in Fig. 1.

VERNIER CALLIPERS

The divisions on the vernier scale are slightly smaller than the main scale divisions. The difference between the value of one main scale division and that of one vernier scale division is constant for a particular vernier. It is called vernier constant or least count of the vernier. It is the least distance which can be accurately measured with the vernier. Therefore, it is also called the least count of the vernier. In general, N vernier divisions are equal to $(N–1)$ main scale divisions.

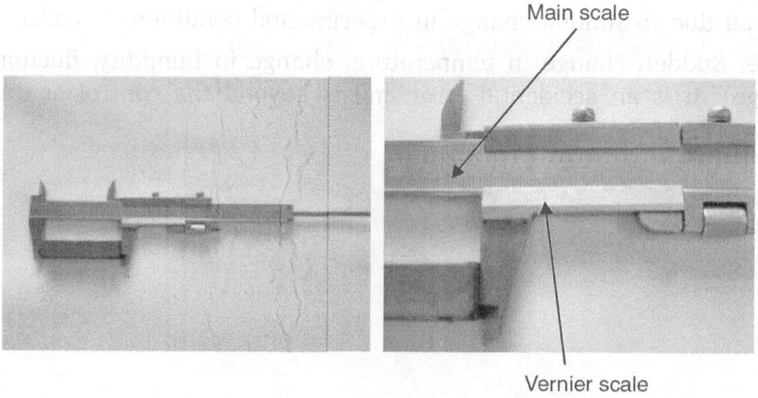

Fig. 2

i.e., $$N \text{ V.S.D.} = (N - 1) \text{ M.S.D.}$$

$$1 \text{ V.S.D.} = \frac{N-1}{N} \text{ M.S.D.}$$

$$= 1 \text{ M.S.D.} - \frac{1}{N} \text{ M.S.D.}$$

or $$1 \text{ M.S.D.} - 1 \text{ V.S.D.} = \frac{1}{N} \text{ M.S.D.}$$

But, $$\text{Least count} = 1 \text{ M.S.D.} - 1 \text{ V.S.D.}$$

$\therefore$ $$\text{Least count} = \frac{1}{N} \text{ M.S.D.}$$

or $$\text{Least count} = \frac{\text{Value of 1 M.S.D}}{\text{Total vernier scale divisions}}.$$

Positive Zero Error. On bringing both the jaws together, if the zero mark of the vernier scale is on the right hand side of the zero mark of the main scale, the error is positive. The value of zero error is obtained by getting the number of division of vernier, coincident with any division of main scale. The number of division of vernier is then multiplied by least count to get the value of zero error in proper unit.

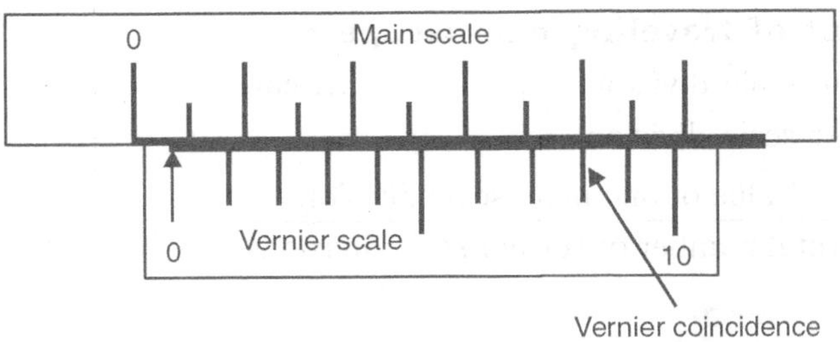

Fig. 3. *Positive error.*

Negative Zero Error. On bringing both the jaws together; if the zero mark of the vernier scale is on the left hand side of the zero mark of the main scale, the error is negative.

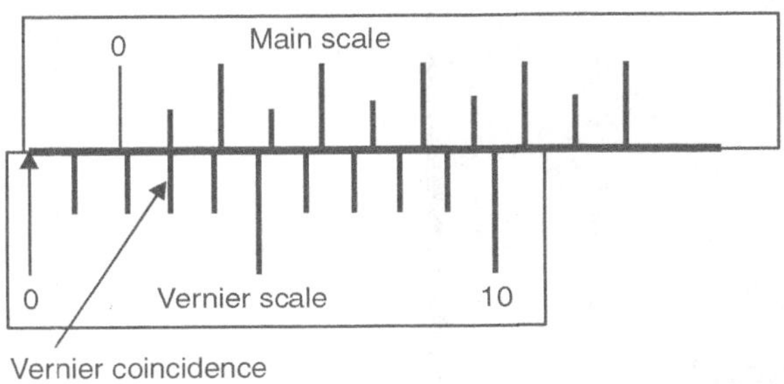

Fig. 4. *Negative error.*

Total Reading, T.R. = M.S.D. + [Vernier Coincidence (V.C.) × Least Count (L.C.)]

Accurate Reading = Total Reading (T.R.) – Zero Error with sign.

> **Note.** *Never estimate the value less than the least count of the measuring instrument.*

TRAVELING MICROSCOPE

The instrument is used to measure small vertical or horizontal distance with high accuracy.

Least Count is the least distance that can be measured by the instrument.

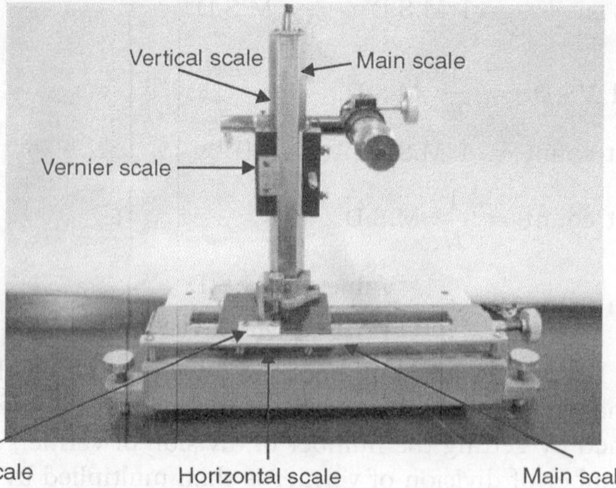

Fig. 5

For Least Count of traveling microscope :

Value of one main scale division = ... cm

Number of vernier scale division = ...

$$\text{Least Count} = \frac{\text{Value of one main scale division}}{\text{Total number of vernier scale divisions}} = \text{..... cm}$$

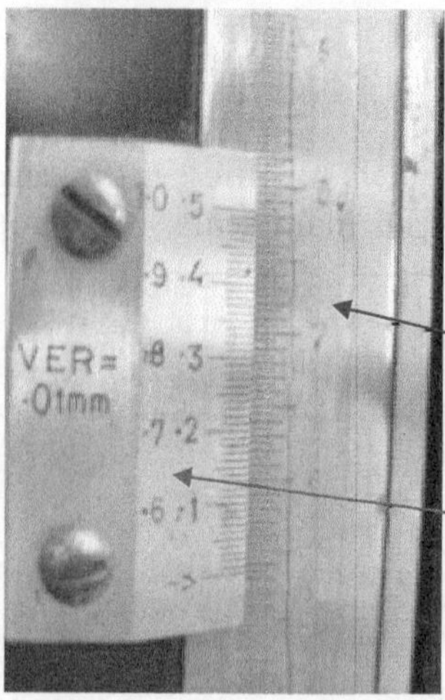

Fig. 6

SPECTROMETER

The same principle applies to circular vernier used for the measurement of angles in spectrometers.

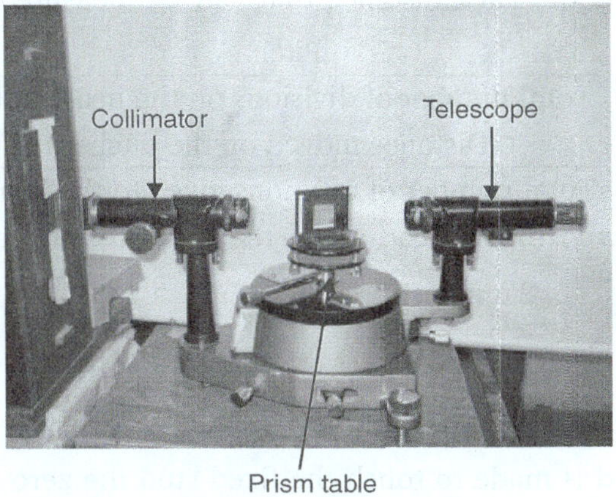

Fig. 7

Least Count of the spectrometer :

$$\text{Least Count} = \frac{1\,\text{M.S.D.}}{\text{Total no. of V.S.D.}} = \frac{0.5^\circ}{30} = \frac{30'}{30} = 1'$$

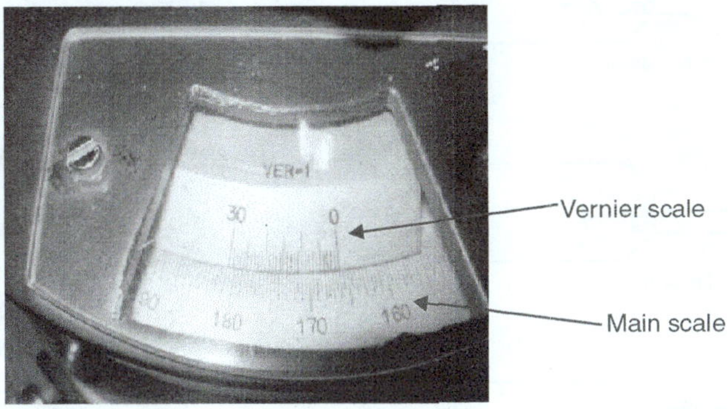

Fig. 8

SCREW GAUGE

Purpose of using a screw gauge is to find the thickness of thin (small) objects.

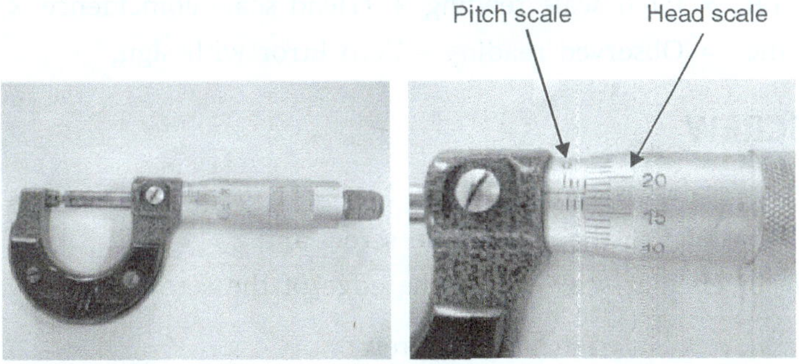

Fig. 9

Least Count : This is the least distance that can be measured by the instrument.

$$\text{Least Count} = \frac{\text{Pitch}}{\text{Total number of divisions on the head scale}}$$

$$\text{Pitch} = \frac{\text{Distance moved on the pitch scale}}{\text{Total number of rotations given to the head scale}}$$

If two rotations are given to the head scale, the distance moved is 2 mm in pitch scale.

$$\therefore \quad \text{Pitch} = \frac{2\text{ mm}}{2} = 1\text{ mm}$$

$$\therefore \quad \text{Least Count} = \frac{1}{50} = 0.02\text{ mm}$$

When the movable stud is made to touch the fixed stud the zero of the Pitch scale and head scale must coincide. If not the screw gauge is said to possess an error. The value of zero error is obtained by getting the number of divisions of head scale above or below the zero of pitch scale. This number is then multiplied by least count to get the value of zero error.

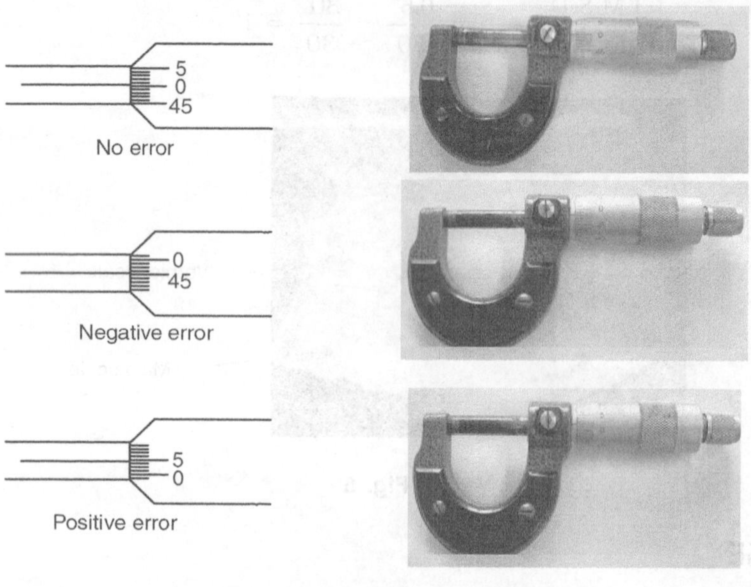

Fig. 10

Total reading, T.R. = Pitch scale reading + (Head scale coincidence × L.C.) in mm

Corrected reading = Observed reading – Zero Error with sign.

MICROMETER SCREW

It is provided with two scales as shown in Fig. 11 :

1. A pitch scale parallel to the axis of the screw and
2. A head scale or circular scale round the edge of the screw head.

$$\text{Least Count} = \frac{\text{Pitch of the screw}}{\text{Total number of divisions on the head scale}}$$

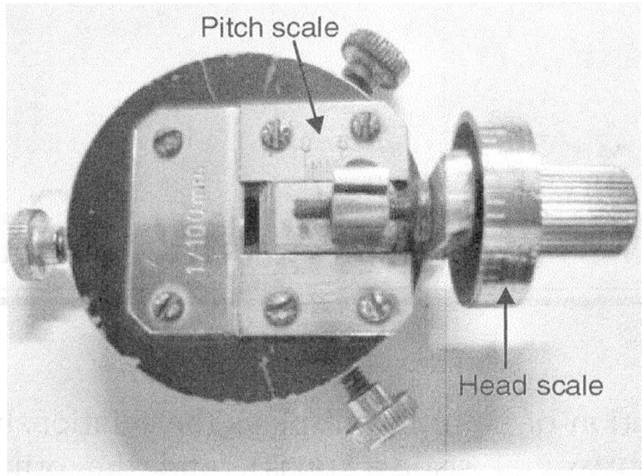

Fig. 11

MICROMETER EYEPIECE

This measuring device is used as a part of the biprism experiment. It contains an eyepiece and pitch and head scale arrangement the eyepiece used here is called a Rams den's eyepiece. It consists of two plano-convex lenses with their spherical surfaces facing each other. The eyepiece also has cross wires. The fringe pattern or any image produced is observed through eyepiece. Pitch and head readings are taken by coinciding images of fringe pattern with the cross-wires.

Fig. 12

Least Count: This is the least distance that can be measured by the instrument.

$$\text{Least Count} = \frac{\text{Pitch}}{\text{Total number of divisions on the head scale}}$$

$$\text{Pitch} = \frac{\text{Distance moved on the pitch scale}}{\text{Total number of rotations given to the head scale}}$$

(If two rotations are given to the head scale, the distance moved is 2 mm in pitch scale.)

$$\therefore \quad \text{Pitch} = \frac{2\text{ mm}}{2} = 1 \text{ mm}$$

$$\therefore \quad \text{Least Count} = \frac{1}{50} = 0.02 \text{ mm}$$

❑❑❑

How to Draw Graph

A graph is a visual representation of data that displays the relationship between two sets of data, one set represented on a horizontal scale (*X*-axis), and the other on a perpendicular scale (*Y*-axis). The results of measurement can be represented more effectively by a graph.

TO DRAW A GRAPH

Draw two mutually perpendicular lines *X'X* and *Y'Y* by a sharp pencil (as shown in Fig. 1). The horizontal line *X'X* is named *X*-axis and vertical line *Y'Y* is *Y*-axis. The point where the two axes meet is the origin. It is named *O* and its co-ordinates are (0, 0).

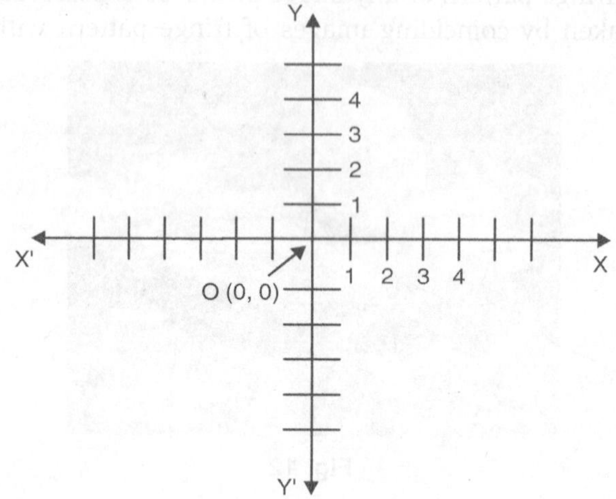

Fig. 1

1. To the left of *O*, *x* values are negative and to the right of it *x* values are positive. Similarly above it *y* values are positive and below it *y* values are negative.
2. In many experiments all the values of both *x* and *y* are positive. So, we require only the first quadrant of the graph.
3. In this case draw *X*-axis at the bottom edge of the graph paper and *Y*-axis near the left edge of the graph paper. So the origin will be at the left hand bottom corner of the graph paper at the intersection of the two axes.
4. Mark origin *O* and axes as *OX* and *OY*.
5. Plot the independent variable (chosen by you) along *X*-axis and the dependent variable along *Y*-axis.

6. Label both the axes with units in bracket. The increasing direction of the scale must be indicated by an arrow. *Scale chosen should be mentioned positively at the upper right corner of graph paper.*
7. Now mark each data on the graph paper by a dot with a sharp pencil in its proper place within the accuracy of one small square according to scale and encircle it.
8. See that each data has been marked.
9. Draw the best smooth curve or straight line through the points; ignore the points that are obviously erratic.
10. But calibration curves are point to point graph.

TO FIND THE SLOPE OF THE STRAIGHT LINE OBTAINED ON THE GRAPH

Plot the graph by following steps. If the graph is a straight line, to find the slope take two points *A* and *B* on the line (Fig. 2), at a distance more than half the length of the line. These points should be so chosen that the values of x and y for these points can be read on the axes conveniently. Draw a horizontal line *AC* from *A* which meets at point *C* on *Y*-axis and *BD* from *B* which meet at point *D* on *Y*-axis. Similarly, draw a vertical line *AE* from *A* which meet at point *E* on *X*-axis and *BF* from *B* which meet at point *F* on *X*-axis.

Let values of x and y corresponding to the points *A* and *B* be (x_1, y_1) and (x_2, y_2), respectively, then the slope of straight line would be equal to $\frac{(y_2 - y_1)}{(x_2 - x_1)}$.

$$\text{The slope of the line } s = \frac{BG}{AG} = \frac{DC}{EF} = \frac{(y_2 - y_1)}{(x_2 - x_1)}$$

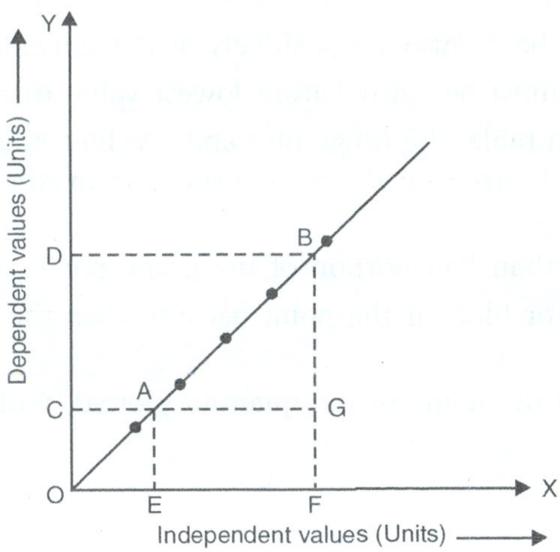

Fig. 2

TO FIND THE SLOPE OF THE CURVE OBTAINED ON THE GRAPH

When the graph is a curve, the slope of the curve cannot be obtained by above method because the gradient is different at different points on the curve. So in this case the slope is calculated corresponding to a chosen point on the curve by following method.

In Fig. 3 a tangent is drawn at a point *A* on the graph. Select other two points *B* and *C* on the tangent. Draw horizontal lines *BD* and *CE* and vertical lines *BF* and *CG* from these points on the axes as shown in Fig. 3.

The slope of the tangent at *A* is the slope of the curve at that point.

Slope at *A*, is $\dfrac{BH}{HC} = \dfrac{DE}{FG}$

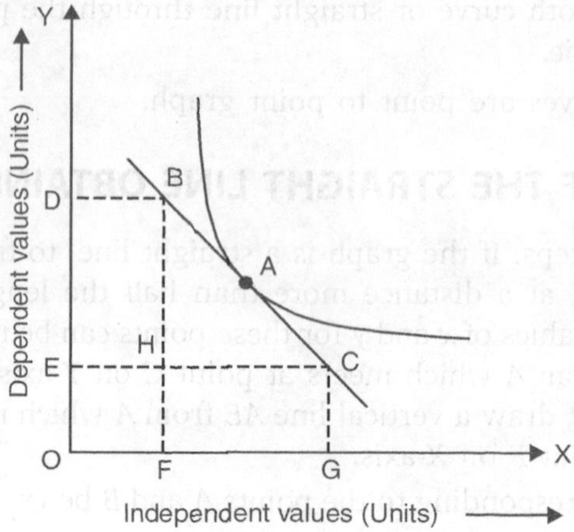

Fig. 3

PRECAUTIONS

1. The scale chosen must be such that it is easily subdivided for the divisions on the graph paper.
2. Scale chosen should be mentioned positively at the upper right corner of graph paper.
3. The variable values must be started from lowest value to highest value at both axes.
4. From the observation table, the range of *x* and *y* values must be chosen such that, these points can be plotted easily on the axes, using the maximum part of the graph paper provided.
5. In every case, more than half portion of the graph paper must be used.
6. Do not make a blot or blots at the point because then the location of exact data point becomes difficult.
7. Label each axis with the name of the quantity plotted, with its units.

❑❑❑

INTERNAL EVALUATION CHART FOR TEACHERS

Internal Evaluation Chart for Teachers

Section

Session

Class Roll No.	Name of the Student	Turn 1 Previous Work Completed	Turn 1 Present Readings Taken	Turn 1 Integrity	Turn 2 Previous Work Completed	Turn 2 Present Readings Taken	Turn 2 Integrity	Up to....	Turn 10 Previous Work Completed	Turn 10 Present Readings Taken	Turn 10 Integrity

EXPERIMENTS

EXPERIMENT—01

AIM

Calibration of a voltmeter with a potentiometer.

APPARATUS REQUIRED

Potentiometer, two storage batteries, two rheostats, galvanometer, a standard cell, two one way keys, one two way key, a Voltmeter (0–1 V)

Fig. 1.1. *Experimental set-up with potentiometer.*

DESCRIPTION OF APPARATUS

The potentiometer described here is the stretched wire potentiometer. It is made up of a long wire usually having a length of 5 m or 10 m. This wire is fixed on a wooden platform and has a uniform diameter. Platform being only 1 m long the separate wires of 1 m lengths are fixed parallel to each other on the platform. These wires are joined to each other by thick copper strip so as the combination acts as a single wire. A driving circuit is connected between the two ends *A* and *B* of the potentiometer, as shown in Fig. 1.2. This circuit consists of a series connection of a strong battery, a plug key and a rheostat, and provides a constant current through the wire *AB*. A scale is fitted on the platform parallel to the wire to measure the distance of any point of wire from one end point, *A*.

Any moving coil or moving magnet voltmeter measuring the potential difference across two points of a electrical circuit draws a small current, and hence changes the potential difference between the two points but the potentiometer measures the potential difference by drawing no current from the circuit and hence it may be considered to be an ideal voltmeter having infinite resistance.

PRINCIPLE

If the two cells are connected as shown in Fig. 1.1, flow of current through the galvanometer can be determined based on the following three conditions :

1. If EMF of cell 1 is greater than EMF of cell 2 current flows in one direction.
2. If EMF of cell 1 is less than the EMF of cell 2 current flows in other direction.
3. If EMF of cell 1 becomes exactly equal to EMF of cell 2 no current flows through the galvanometer.

So, if the potential difference across two points of a circuit is equal to the potential drop across the specific length of potentiometer wire than no current flows through the galvanometer and the potential drop can be found by knowing the potential drop across the length of potentiometer wire.

THEORY

The resistance of a wire is measured by

$$R = \rho\frac{L}{A}$$

where R is the total resistance,

ρ is the resistivity of the metal of a wire at a definite temperature,

L is the length of a wire,

and A is the area of cross-section.

If the wire is of uniform diameter and we consider its unit length say of 1 cm, then the resistance

$$R = \frac{\rho}{A} \qquad \text{(This is a constant.)}$$

Therefore, we may consider the wire of 10 m as the series combination of 1000 resistances each having a resistance of $\frac{\rho}{A}$ Ω. When a constant current is flowing through the potentiometer wire it creates a constant drop across each unit length of wire and each length of wire becomes a calibrated measurement of potential. If a total potential of V_0 drops across the total length L of wire then the drop per unit length is given by

$$k = \frac{V_0}{L}$$

And hence any length l of the potentiometer between the point A and P (Fig. 1.2) shall be having a potential of

$$V' = kl$$

$$V' = V_0\frac{l}{L}$$

FORMULA USED

The error in voltmeter reading is given by

$$V' - V = V_0 \frac{l}{L} - V$$

where V = Potential difference between two points read by voltmeter,

V' = Potential difference between the same two points read by potentiometer,

V_0 = EMF of the standard cell,

L = Length of the potentiometer wire corresponding to E.M.F. of the standard cell,

l = Length of the potentiometer wire corresponding to the potential difference V', measured by potentiometer.

EXPERIMENTAL ARRANGEMENT

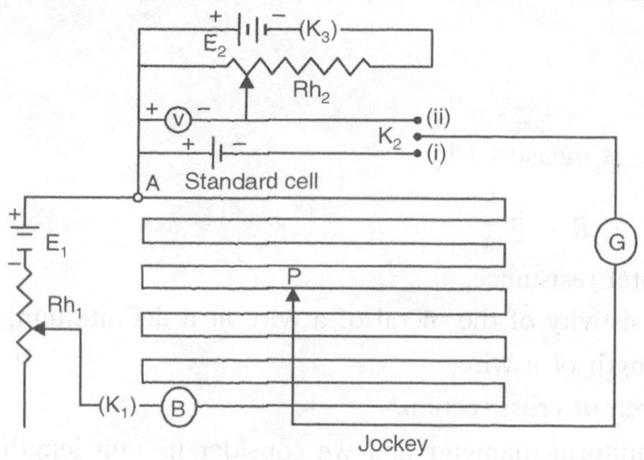

Fig. 1.2. *Experimental arrangement.*

PROCEDURE

For Calibration of Potentiometer Wire

1. Make the electrical connections as shown in Fig. 1.2.
2. Place the jockey on the last end of the wire so that total length of the wire is 1000 cm. Close K_1 and $K_2(i)$. In this case a deflection is observed in the galvanometer, which reduced to zero by adjusting the rheostat Rh_1 in the main circuit. At this stage the potential drop across the potentiometer wire is exactly equal to the *EMF* of the standard cell.
3. Note down the total balancing length L (which have been taken as 1000 cm) and EMF (V_0) of the standard cell. The adjustment of the rheostat Rh_1 should not be disturbed in performing the next steps.

For Calibration of Voltmeter

4. Open $K_2(i)$ and close $K_2(ii)$. By closing K_3, adjust the Rheostat Rh_2 (the potential divider) such that the voltmeter reading may be 0.1 volt. Adjust the jockey on the wire such that there is no deflection in the galvanometer. Note down the total balancing length l.

5. By readjusting the Rheostat Rh_2 and getting a new reading in the voltmeter repeat the step 4, till the entire range of the voltmeter is covered.
6. Get the calculated value of potential differences (V'). Plot the graph between the observed value V of the voltmeter and the error ($V' - V$). This graph is a point-to-point graph.

OBSERVATIONS

Table 1.1

Length of the potentiometer wire corresponding to EMF of standard cell L (cm)	***Remarks***
	EMF of standard cell, V_0 = ... volts Potential gradient, $k = \frac{V_0}{L}$ = ... volts/cm

Table 1.2. For the calibration of potentiometer wire.

S.No.	***Voltmeter reading V (volts)***	***Balancing length l (cm)***	***Accurate P. D. V′ = kl (volts)***	***Error V′ – V (volts)***
1.				
2.				
3.				
4.				
5.				
6.				
7.				
8.				
9.				
10.				

CALCULATIONS

The potential gradient is calculated by the following formula:

$$k = \frac{V_0}{L} \text{ volts per unit length}$$

The Correct value of potential is calculated by the formula

$$V' - V = V_0 \frac{l}{L} - V$$

Similarly, calculate the error for other readings. Draw a graph between the error $(V' - V)$ and the voltmeter reading. The shape will be zigzag as shown in Fig. 1.3.

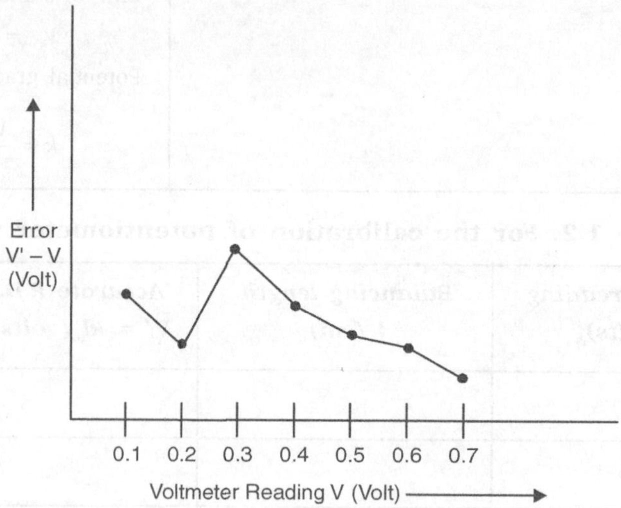

Fig. 1.3. *Graph.*

RESULT

The plotted graph between the error $(V' - V)$ and the voltmeter reading V is the calibration curve of the given voltmeter.

PRECAUTIONS

1. EMF of E_1 should be greater than that of E_2 and standard cell.
2. Potential difference between A and B should be greater than the maximum potential difference to be measured. EMF of E_2 should be greater than the range of voltmeter to be calibrated.
3. Plug keys K_1, K_2 should be used only when observations are to be taken.
4. K_3 should be kept open when potential difference across rheostat Rh_2 is measured by potentiometer.

CONCLUSION

VIVAVOCE

1. What is a potentiometer ?
2. Why are you calibrating your voltmeter ?
3. Why did you take L = 950 cm or 1000 cm ? Why not 500 cm ?
4. Potentiometer wire is made of constantan and not of copper or iron why ?
5. A voltmeter takes some current in measuring. Is it not error due to this ?
6. Do you get balancing at null point by the galvanometer ?
7. What do you mean by calibration ?
8. What do you mean by resistance of a wire ?
9. What do you mean by potential gradient ?
10. What is the resistance of potentiometer wire ?

❑❑❑

EXPERIMENT—02

AIM

To compare the electromotive forces (e.m.f.) of two given cells by Lumsden's method.

APPARATUS REQUIRED

Two cells (a Daniel cell and a Leclanche cell or its electronic equivalent), a tapping key, two resistance boxes and a galvanometer.

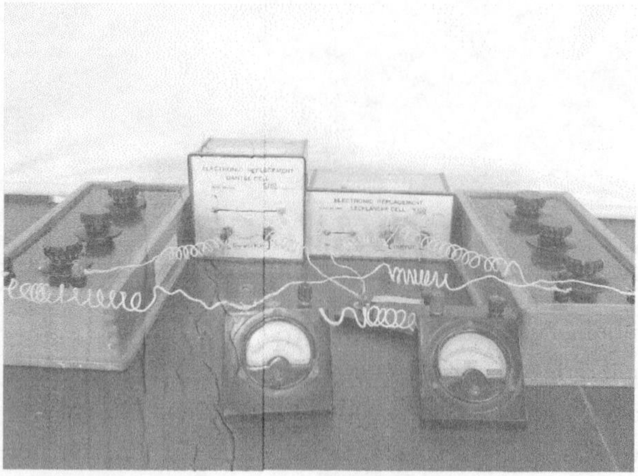

Fig. 2.1. *Setup for Lumsden's method.*

DESCRIPTION OF THE APPARATUS

The apparatus consists of two cells namely Daniel cell and Leclanche cell.

Daniel Cell

It is the simplest form of a primary cell. It consists of a zinc electrode in a dilute sulphuric acid (H_2SO_4) solution and a copper electrode in copper sulphate ($CuSO_4$) solution. A porous cup separates the two solutions. The zinc electrode works as the negative electrode and the copper electrode works as the positive electrode.

In the cell following reaction takes place :

$$Zn + H_2SO_4 \longrightarrow ZnSO_4 + 2H^+ + 2e^-$$

The electrons released in this reaction are collected at zinc rod and makes it at negative potential and the hydrogen ions interact with copper sulphate and forms sulphuric acid and releases copper ions (Cu^{++}) as shown in following reaction :

$$2H^+ + CuSO_4 \longrightarrow H_2SO_4 + Cu^{++}$$

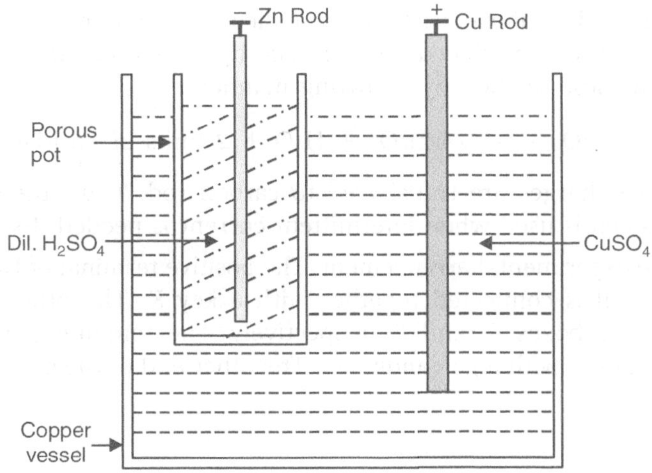

Fig. 2.2. *Daniel cell.*

These copper ions move towards the copper electrode and deposited on it. Thus, the positive charge is given up to this electrode in the process. Thus, an e.m.f. is produced.

Daniel cell is used when a continuous current is required. The e.m.f. of a Daniel cell is around 1.09 V.

Leclanche Cell

It consists of a glass vessel containing strong solution of ammonium chloride (NH_4Cl). An amalgamated zinc rod dipped in ammonium chloride acts as cathode. A porous pot is placed inside the vessel. A carbon rod placed inside the porous pot acts as anode. The space in the porous pot is filled with manganese dioxide (MnO_2) and charcoal powder. The following reaction takes place in the cell :

$$Zn + 2NH_4Cl \longrightarrow 2NH_3 + ZnCl_2 + 2H^{++} + 2e^-$$

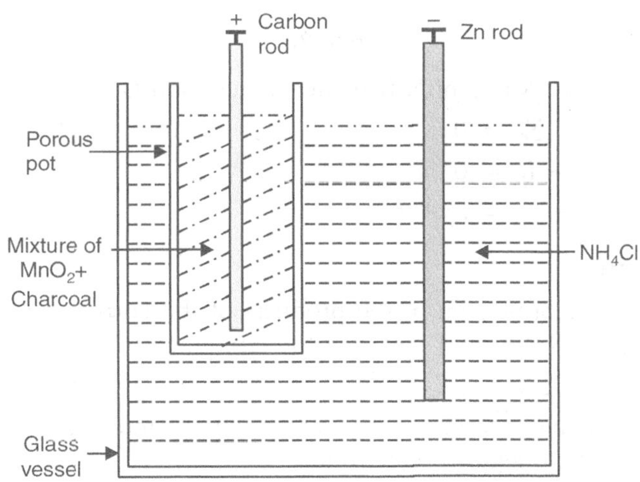

Fig. 2.3. *Leclanche cell.*

The electrons released in above reaction are collected at zinc rod and makes it at negative potential. The ammonia gas so produced escapes. The hydrogen ions diffuse through porous pot and interact with manganese dioxide by following manner :

$$2H^{++} + 2MnO_2 \longrightarrow Mn_2O_3 + H_2O + 2 \text{ units of positive charges}$$

These two positive charges are transferred to carbon rod. Thus, the carbon rod attains at positive potential. This cell is used when intermittent current is needed. Its e.m.f. is about 1.45 V.

Fig. 2.1 shows the experimental arrangement. The positive terminal of Daniel cell and negative terminal of Leclanche cell is connected together with a key *K*. The other ends of the cells are connected with resistance boxes R_1 and R_2 respectively. The remaining terminals of resistance boxes are connected together with a galvanometer. The other end of the galvanometer is connected with key.

PRINCIPLE

The electromotive force (e.m.f.) of a cell is the maximum potential difference between the two electrodes of the given cell when no current is drawn from the cell or cell is in the open circuit. In this method, the e.m.f. of two cells can be compared by applying Kirchhoff's laws, viz.

Kirchhoff's First Law

It states that the algebraic sum of the currents meeting at a junction in a closed circuit is zero.

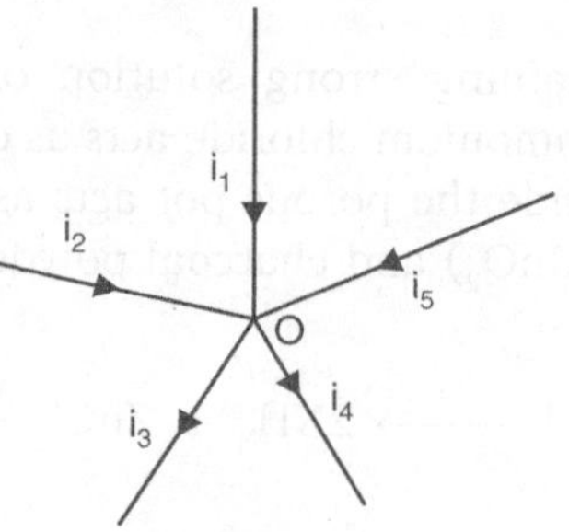

Fig. 2.4

In Fig. 2.4, according to Kirchhoff's first law, at junction *O*

$$\Sigma i = 0$$

$$i_1 + i_2 - i_3 - i_4 + i_5 = 0$$

or

$$i_1 + i_2 + i_5 = i_3 + i_4$$

Kirchhoff's Second Law

In any closed circuit the algebraic sum of the products of the current and the resistance of each part of the circuit is equal to the total e.m.f. in the circuit,

$$\Sigma iR = \Sigma E$$

The positive value of the current is taken when we are moving in the direction of current and e.m.f. is taken positive to negative electrode.

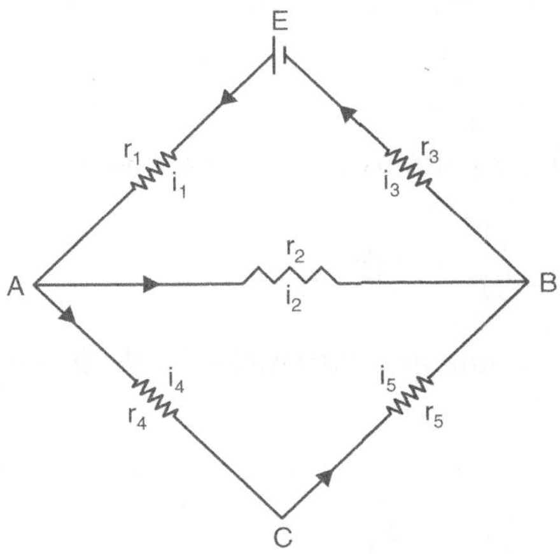

Fig. 2.5

Consider the circuit *EABE* in Fig. 2.5

$$i_1r_1 + i_2r_2 + i_3r_3 + ir = E$$

and in the circuit *ABCA*

$$i_2r_2 - i_5r_5 - i_4r_4 = 0$$

THEORY

Let E_1 and E_2 be the e.m.f.'s of two given cells of internal resistances r_1 and r_2 respectively. R_1 and R_2 are the external resistances applied from resistance boxes. *G* is the resistance of the galvanometer.

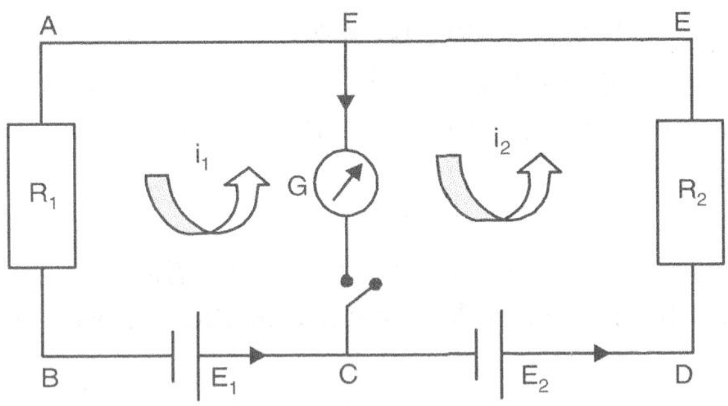

Fig. 2.6. *Circuit diagram for Lumsden's method.*

Let i_1 and i_2 be the cyclic current in the meshes *ABCFA* and *CDEFC* respectively. Now applying Kirchhoff's second law to both meshes, we get

$$(R_1 + r_1)i_1 + G(i_1 - i_2) = E_1 \quad \ldots(1)$$

and

$$(R_2 + r_2)i_2 + G(i_2 - i_1) = E_2 \quad \ldots(2)$$

In equilibrium condition there will be no deflection in the galvanometer, in this case

$$i_1 = i_2$$

or

$$i_1 - i_2 = 0$$

Substituting these values in equations (1) and (2), we get

$$(R_1 + r_1)i_1 = E_1 \quad ...(3)$$

and

$$(R_2 + r_2)i_2 = E_2 \quad ...(4)$$

Taking $i_1 = i_2$ and dividing equations (3) and (4), we get

$$\frac{(R_1 + r_1)}{(R_2 + r_2)} = \frac{E_1}{E_2} \quad ...(5)$$

To eliminate r_1 and r_2, take another observation with different values of R_1 and R_2 for no deflection in the galvanometer.

Then, we get

$$\frac{(R_1' + r_1)}{(R_2' + r_2)} = \frac{E_1}{E_2} \quad ...(6)$$

From equations (5) and (6), we get

$$\frac{(R_1 + r_1) - (R_1' + r_1)}{(R_2 + r_2) - (R_2' + r_2)} = \frac{E_1}{E_2}$$

or

$$\frac{R_1 - R_1'}{R_2 - R_2'} = \frac{E_1}{E_2} \quad ...(7)$$

PROCEDURE

1. Make the connections as shown in Fig. 2.6.
2. Take out the zinc rod of the Daniel cell E_1 *i.e.*, switching off the Daniel cell. Press the tapping key K and note the direction of deflection in the galvanometer. Insert the rod in the cell.
3. Next take out the zinc rod of the Leclanche cell E_2 *i.e.*, switching off the Leclanche cell. Press the key K and again note the direction of deflection in the galvanometer. Again insert the zinc rod. If the two deflections are in the opposite directions, the connections are correct.
4. Take suitable values of R_1 and find out the value of R_2 for which there is no deflection in the galvanometer.
5. Take other value of R_1 and repeat the process (4). Take six or seven readings.
6. Measure the value of E_1 and E_2 with the help of voltmeter.
7. Ratio between E_1 and E_2 may also be calculated by plotting a graph between R_1 and R_2 for their various values.

OBSERVATION TABLE

Table 2.1. For R_1 and R_2.

S.No.	R_1 *(in Ω)*	R_2 *(in Ω)*	$\frac{E_1}{E_2} = \frac{(R_1 - R_1')}{(R_2 - R_2')}$

Mean $\frac{E_1}{E_2}$ =

Table 2.2. For E_1 and E_2 by voltmeter.

EMF of Daniel cell (E_1) (in volts)	*EMF of Leclanche cell (E_2) (in volts)*	$\frac{E_1}{E_2}$

CALCULATIONS

Using the formula, calculate the value of $\frac{E_1}{E_2}$

$$\frac{E_1}{E_2} = \frac{(R_1 - R_1')}{(R_2 - R_2')}$$

GRAPHICAL CALCULATION

Values of R_1 are taken along the *X*-axis and that of R_2 along *Y*-axis and two points *P* and *Q* are taken on the graph. *PM* and *QN* perpendiculars are drawn on the *X*-axis. From the point *Q*, *QR* is drawn perpendicular to *PM*.

It is clear from the graph

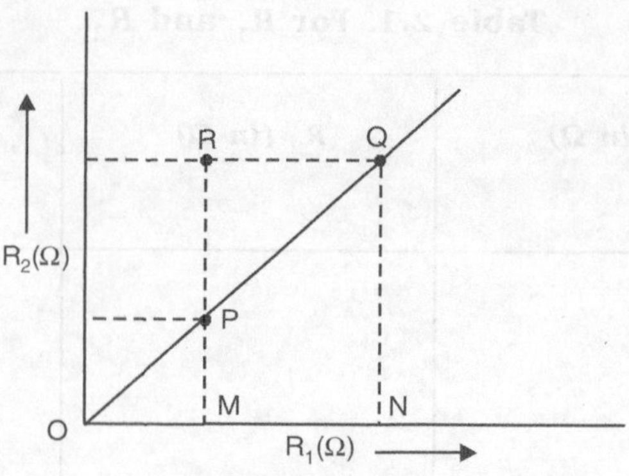

Fig. 2.7. *Graph.*

$$QR = OM - ON$$
$$= R_1 - R_1'$$

and $$PR = PM - RM$$

So, we have $$= R_2 - R_2'$$

$$\frac{E_1}{E_2} = \frac{QR}{PR} = \frac{(R_1 - R_1')}{(R_2 - R_2')}$$

RESULT

1. $\frac{E_1}{E_2}$ by calculation = ...
2. $\frac{E_1}{E_2}$ by graph = ...
3. $\frac{E_1}{E_2}$ by voltmeter = ...

PRECAUTIONS

1. The positive of one cell and the negative of another cell should be connected to the same terminal of the galvanometer.
2. To avoid any damage to the galvanometer it must be properly shunted.
3. A tapping key should always be used in the galvanometer circuit.
4. The values of R_1 and R_2 preferably, should be in hundreds of ohms.

CONCLUSION

VIVAVOCE

1. What is the object of your experiment ?
2. What is a cell ?
3. What do you mean by electromotive force of a cell ?
4. What is the difference between electromotive force and potential difference ?
5. How many types of the cell are there ?
6. What is the difference between primary and secondary cell ?
7. Which types of the cell are used in your experiment ?
8. Explain internal resistance of the cell.
9. How do you eliminate the error due to internal resistance of the cell ?
10. Explain Kirchhoff's laws.
11. Is this method superior to potentiometer method of comparing e.m.f of two cells ?
12. What do you understand by the terms : potential, potential at any point and potential difference between two points ?
13. Explain the construction of any cell you are using.

❑❑❑

EXPERIMENT—03

AIM

To determine :

(*i*) The resistance per unit length of the Carey-foster's bridge wire.

(*ii*) The resistance of experimental wire.

(*iii*) The specific resistance of the experimental wire by comparing the resistance of wire with standard resistance.

Fig. 3.1. *Carey-foster's bridge apparatus.*

APPARATUS REQUIRED

Carey-foster's Bridge, Decimal resistance box, Primary cell, Galvanometer, Thick copper strip, connecting wires, plug key, Rheostat, Jockey and experimental wire.

DESCRIPTION OF THE APPARATUS

The experimental arrangement of Carey-foster bridge is shown in Fig. 3.2. The arrangement is similar to Wheatstone bridge. A wire *AB* of mangenin one meter long, having uniform area of cross-section is fixed on a wooden box. A meter scale is also fixed parallel to the wire. The ends *A* and *B* are fixed with two thick copper strips. Three more copper strips are fixed parallel to the wire. They provide four gaps (1), (2), (3) and (4). A resistance box R.B. and a thick copper strip are connected in gap (1) and gap (2) respectively. A battery *E* is connected across points *C* and *F*. A rheostat (Rh) is also connected to maintain nearly equal resistance *P* and *Q* whose actual values need not to be known. One end of a galvanometer *G* is connected to point *D* and other end act as a jockey *J*. The jockey can be made to slide over wire *AB*.

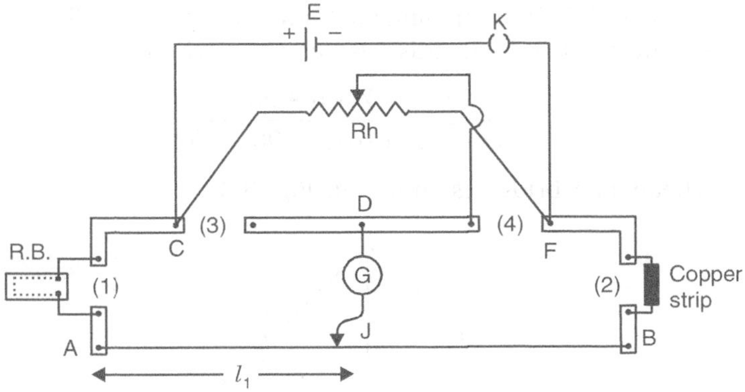

Fig. 3.2. *Carey-foster bridge arrangement.*

PRINCIPLE

Carey-foster Bridge is a modified form of a meter bridge. In this bridge, the length of the bridge wire increases by including two extra resistances in series with the wire. It is thus suitable for accurate comparison of the nearly equal resistances.

THEORY

Let ρ be the resistance per unit length of the bridge wire and α and β are end resistances. A resistance box R.B. is connected in left gap (1) and an experimental wire of resistance Y is connected in right gap (2). If X is the resistance offered by resistance box R.B. and the null point D is obtained at the distance l_1 from point A. Then in this condition, according to Wheatstone Bridge Principle

$$\frac{P}{Q} = \frac{X + \alpha + \rho l_1}{Y + \beta + \rho(100 - l_1)} \qquad \ldots(1)$$

The equivalent Wheatstone bridge is shown in Fig. 3.3.

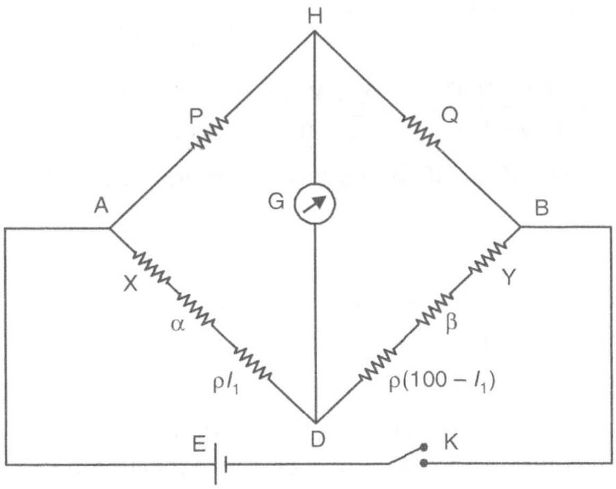

Fig. 3.3

If the position of X and Y is interchanged then position of null point from A will also changed. If position of null point is l_2. In this case at equilibrium

$$\frac{P}{Q} = \frac{Y + \alpha + \rho l_2}{X + \beta + \rho\,(100 - l_2)} \qquad \ldots(2)$$

The equivalent Wheatstone bridge is shown in Fig. 3.4

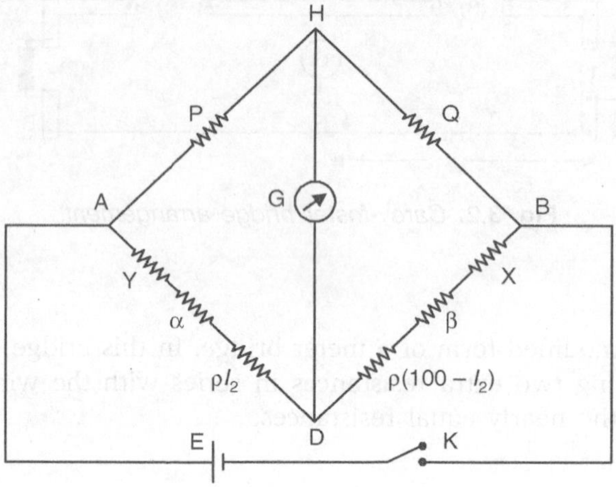

Fig. 3.4

From above two equations, we get

$$\frac{X + \alpha + \rho l_1}{Y + \beta + \rho\,(100 - l_1)} = \frac{Y + \alpha + \rho l_2}{X + \beta + \rho\,(100 - l_2)}$$

Adding 1 to both sides, we obtain

$$\frac{X + \alpha + \rho l_1}{Y + \beta + \rho\,(100 - l_1)} + 1 = \frac{Y + \alpha + \rho l_2}{X + \beta + \rho\,(100 - l_2)} + 1$$

$$\frac{X + \alpha + \rho l_1 + Y + \beta + \rho\,(100 - l_1)}{Y + \beta + \rho\,(100 - l_1)} = \frac{Y + \alpha + \rho l_2 + X + \beta + \rho\,(100 - l_2)}{X + \beta + \rho\,(100 - l_2)}$$

$$\frac{X + Y + \alpha + \beta + 100\rho}{Y + \beta + \rho\,(100 - l_1)} = \frac{X + Y + \alpha + \beta + 100\rho}{X + \beta + \rho\,(100 - l_2)}$$

Since numerators on both sides of the equation are equal, hence the denominator must also be equal.

Therefore, we get

$$Y + \beta + \rho(100 - l_1) = X + \beta + \rho(100 - l_2)$$

$$X - Y = \rho(l_2 - l_1) \qquad \ldots(3)$$

$$Y = X - \rho(l_2 - l_1) \qquad \ldots(4)$$

Thus, the difference between two resistances is equal to the resistance of the bridge wire between the two balance points and resistance of the experimental wire can be calculated by equation (4). If the experimental wire is replaced by a thick copper strip then, from equation (3).

$$X - Y = \rho(l_2 - l_1)$$

Assuming $Y = 0$ (For copper strip)

$$X = \rho(l_2 - l_1)$$

$$\rho = \frac{X}{(l_2 - l_1)} \qquad \text{...(5)}$$

FORMULA USED

1. Resistance per unit length of the bridge wire

$$\rho = \frac{X}{(l_2 - l_1)}$$

2. Resistance of the experimental wire

$$Y = X - \rho(l_2' - l_1')$$

3. Specific resistance of the experimental wire

$$S = Y\frac{\pi r^2}{L}$$

where,
ρ = Resistance per unit length of bridge wire.
X = Known resistance of resistance box when copper strip is connected at one end.
l_1 = Length of balance point from A before interchanging R.B. and copper strip.
l_2 = Length of balance point from A after interchanging R.B. and copper strip.
l_1' = Length of balance point from A before interchanging R.B. and unknown resistance wire.
l_2' = Length of balance point from A after interchanging R.B. and unknown resistance wire.
Y = Unknown resistance of the experimental wire.
L = Length of given wire.
r = Radius of given wire.

PROCEDURE

I. For resistance per unit length of bridge wire

1. Make connections as shown in Fig. 3.2.
2. Keep zero resistance in the resistance box; put the jockey at 50 cm mark on the bridge wire. Closing the key K, move the slider of the rheostat sideways till there is no deflection in the galvanometer. Do not touch the rheostat for the remaining part of the experiment.
3. Introduce some resistance X in the ohm box. Obtain the null point by moving jockey on the wire and note the distance l_1 of the null point from left end.
4. Now interchange the positions of the copper strip and resistance box. Obtain the null point by moving jockey on the wire and note the distance l_2 of the null point from left end.
5. Repeat step (3) and (4) for different values of X and note the corresponding values of l_1 and l_2.

II. For the resistance of experimental wire

1. Replace copper strip by experimental wire.
2. Introduce some resistance X in the ohm box. Obtain the null point by moving jockey on the wire and note the distance l_1' of the null point from left end.
3. Now interchange the positions of the experimental wire and resistance box. Obtain the null point by moving jockey on the wire and note the distance l_2' of the null point from left end.
4. Repeat step (2) and (3) for different values of X and note the corresponding values of l_1' and l_2'.
5. All the ten readings should be taken.

OBSERVATIONS

Table 3.1. For resistance per unit length of bridge wire (ρ).

S.No.	*Resistance in the resistance box 'X' (in Ohms)*	*Position of null point with the R.B. in*		$(l_2 - l_1)$	$\frac{X}{(l_2 - l_1)}$	*Mean ρ*
		left gap (l_1) *(cm)*	*right gap* (l_2) *(cm)*	*(cm)*	*(Ohm/cm)*	*(Ohm/cm)*
1.						
2.						
3.						
4.						
5.						
6.						
7.						
8.						
9.						
10.						

Mean ρ = ...

Table 3.2. For the resistance of experimental wire (Y).

S.No.	*Resistance in the resistance box 'X' (in Ohms)*	*Position point of null with R.B. in the*		$(l_2' - l_1')$	$Y = X - \rho(l_2' - l_1')$	*Mean Y*
		Left gap (l_1') *(cm)*	*Right gap* (l_2') *(cm)*	*(cm)*	*(Ohms)*	*(Ohms)*
1.						
2.						
3.						
4.						
5.						
6.						
7.						
8.						
9.						
10.						

Length of the wire = ... cm

Radius of the experimental wire :

$$\text{Pitch of screw gauge} = \frac{\text{Distance travelled on the pitch scale}}{\text{Total number of rotations of circular head}} = \text{..............} \text{ cm}$$

(If 2 rotations are given to the circular scale, the distance moved is 1 mm, the Pitch

$$= \frac{1}{2} = 0.5 \text{ mm.)}$$

No. of circular scale divisions = ...

$$\text{Least Count of screw gauge} = \frac{\text{Pitch}}{\text{Total number of divisions on the circular scale}} = \text{.........} \text{ cm}$$

Zero error = ± .... cm

I. If the zero of the C.S. lies below the base line of M.S. the error is positive.

II. If the zero of the C.S. lies above the base line of M.S. the error is negative.

Table 3.3. For diameter of the experimental wire.

Diameter along any direction			*Diameter along perpendicular direction*			*Mean diameter*	*Mean radius*
Main scale reading	*Circular scale reading*	*Total reading (A) (cm)*	*Main scale reading*	*Circular scale reading*	*Total reading (B) (cm)*	$d = \left(\frac{A+B}{2}\right)$ *(cm)*	$r = \frac{d}{2}$ *(cm)*

CALCULATIONS

1. For the resistance per unit length of the bridge wire :

From observation Table 3.1, the resistance per unit length of the bridge wire is calculated by using the following relation :

$$\rho = \frac{X}{(l_2 - l_1)}$$

Mean ρ = ... Ohm/cm

2. For the resistance of experimental wire :

From observation Table 3.2, the resistance of the experimental wire is calculated by using the following relation :

$$Y = X - \rho(l_2' - l_1')$$

Y = ... Ohm

3. For the specific resistance of experimental wire :

From observation Table 3.3, the specific resistance of experimental wire is calculated by using the following relation :

$$S = Y\frac{\pi r^2}{L}$$

S = ... Ohm.cm

RESULT

1. The resistance per unit length of the bridge wire ρ = ... Ohm/cm
2. The resistance of the experimental wire is Y = ... Ohm
3. The specific resistance of the material of the experimental wire S = ... Ohm/cm

PRECAUTIONS

1. The plugs of fractional resistance box should be tight.
2. Rheostat once adjusted should not be disturbed throughout the experiment.

3. Jockey should be pressed momentarily and should not be pressed when it is moved on the wire.
4. The cell circuit must be completed before the galvanometer circuit.
5. The battery key should be taken out while interchanging X and Y to avoid heating of bridge wire.
6. While determining the value of ρ, the value of standard resistance should be smaller than the resistance of the bridge wire; otherwise null point shall not be obtained.

CONCLUSION

VIVAVOCE

1. What is the object of your experiment ?
2. What is resistance ?
3. Define specific resistance.
4. What are its units ?
5. How does specific resistance vary with temperature ?
6. On which principle Carey-foster bridge works ?
7. Explain null point.
8. Explain the principle of Wheatstone bridge.
9. Explain Kirchoff's laws.
10. What is Meter Bridge ?
11. How can you differ Carey-foster bridge from Meter Bridge ?
12. What is the advantage of Carey-foster bridge over Meter Bridge ?
13. What is a galvanometer ? What is the principle of working of a moving coil type of galvanometer ?
14. What is the difference in the use of a resistance box and a rheostat ?
15. What is the difference between primary and secondary cells ?

❑❑❑

EXPERIMENT—04

AIM

To determine the frequency of A.C. mains by means of a Sonometer.

APPARATUS REQUIRED

Sonometer having non-magnetic wire stretched over it, step-down transformer, magnets (either two or one horse shoe), rheostat, 50/100 gm slotted weights.

Fig. 4.1. *Sonometer.*

DESCRIPTION OF THE APPARATUS

The sonometer is shown in Fig. 4.1. It is a sound box, which is a simple wooden box approximately one meter long. There are four bridges on the sound box. One bridge *P* is fixed at one end of the box and the bridge *Q* is made up of a frictionless pulley. There are two more knife edged bridges *R* and *S* which may slide on the box under the experimental wire. One end of the experimental wire is fixed at the first bridge *P* and the second end passes over the frictionless pulley *Q* and holds the weight holder, on which suitable weight can be put to apply required tension.

The wire between the two sliding bridges *P* and *Q* constitutes the **"String fixed at both ends".** The length of the wire can be changed by varying the distance between the two sliding knife edged bridges *R* and *S*. The tension on the wire can be changed by changing the weight on the weight holder. The vibrations on the experimental wire can be produced either mechanically or electromagnetically depending on the experiment.

PRINCIPLE

The frequency of A.C. mains can be determined by indirect method, by knowing the natural frequency of a piece of wire fixed on both ends. A stretched wire carrying current is when subjected to a magnetic field perpendicular to the direction of current; a force acts in a direction perpendicular to both current and the magnetic field. The direction of the force is decided by the left hand rule. As the current flowing through the wire is alternating, the direction of the force acting on wire changes continuously and the frequency of change is equal to the frequency of the alternating current. If the length of the wire between the bridges is such that its natural frequency is equals the frequency of A.C. mains the resonance occurs and the amplitude of vibration becomes maximum.

THEORY

Consider a sonometer wire is stretched under a constant load and carrying alternating current is placed inside a uniform magnetic field acting in the horizontal plane. Due to the applied magnetic field a force act on the wire in upward direction for positive half cycle and in downward direction for the negative half cycle. Because of this force the wire will move in upward direction for one half cycle and in the downward direction for the next half cycle of A.C. As a result, the wire will execute a forced vibration and the frequency of vibration will be equal to frequency of alternative current.

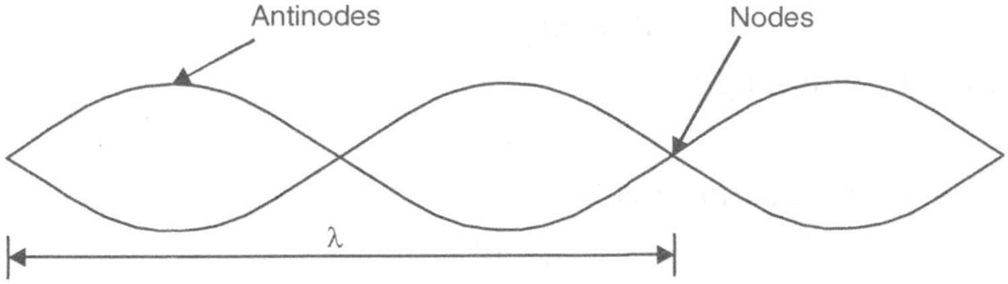

Fig. 4.2. *Stationary waves.*

If the distance between the two movable bridges is so adjusted that wire's natural frequency becomes equal to the frequency of A.C. mains. The Stationary waves (see Fig. 4.2) of maximum amplitude are produced due to resonance. The standing wave will have its nodes at the bridges and an antinode at the centre of the bridges. Thus the wire acts as a stretched string fixed at both the ends. This is called the fundamental mode of vibration (see Fig. 4.3) and in this condition.

$$l = \frac{\lambda}{2}$$

or $$\lambda = 2l \qquad ...(1)$$

where l is the length of the wire between the bridges and λ is the wavelength of the transverse wave.

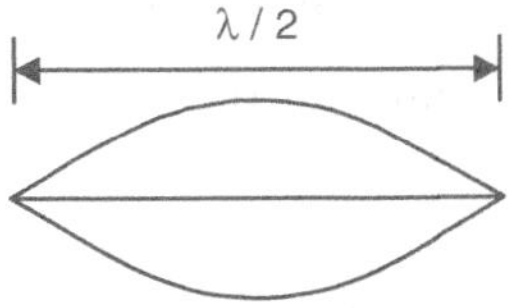

Fig. 4.3. *Fundamental mode of vibration.*

If the frequency of the wave is given by ν and the velocity of the transverse wave is V along the wire then

$$V = \nu\,\lambda$$

Putting the value of λ from equation (1), we get

$$V = 2l\nu \qquad \text{...(2)}$$

The velocity of the transverse wave in the stretched string is equal to the square root of the ratio of the tension to the mass per unit length of wire.

So, $$V = \sqrt{\frac{T}{m}} \qquad \text{...(3)}$$

From equations (2) and (3), we get

$$2l\nu = \sqrt{\frac{T}{m}}$$

or $$\nu = \frac{1}{2l}\sqrt{\frac{T}{m}} \qquad \text{...(4)}$$

If M is the mass suspended on the wire, r be the radius of the sonometer wire and ρ be the density of the material of wire then,

Tension applied on the wire

$$T = Mg$$

and mass of the wire per unit length

$$m = \pi r^2 \rho$$

Putting these values in equation (4), we get

$$\nu = \frac{1}{2l}\sqrt{\frac{Mg}{\pi r^2 \rho}}.$$

FORMULA

The frequency of transverse vibrations of stretched string is given by the formula

$$\nu = \frac{1}{2l}\sqrt{\frac{Mg}{\pi r^2 \rho}}$$

where ν = Frequency of A.C. mains,

l = Length of sonometer wire between the knife edged bridges at resonance,

g = Acceleration due to gravity,

M = Mass kept at the weight holder,

r = Radius of the sonometer wire,

ρ = Density of the material of sonometer wire.

DIAGRAM

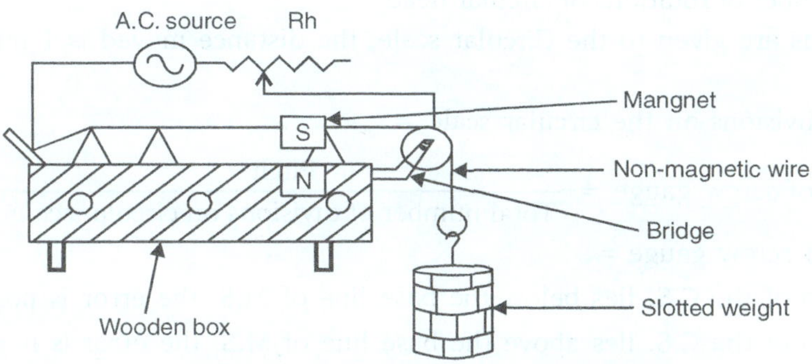

Fig. 4.4. *Experimental arrangements.*

PROCEDURE

1. The connections are made as shown in Fig. 4.4.
2. A weight of 50 (or 100) gm is suspended on one end of the wire.
3. A magnet is mounted horizontally at the center of the wire.
4. The movable bridges are adjusted till the wire starts vibrating with maximum amplitude *i.e.,* till resonance occurs.
5. The length of wire for resonance between the movable bridges is measured.
6. The procedure is repeated by increasing the applied weight gradually and the values of length '*l*' are noted accordingly.
7. When the highest weight is applied, the weights are decreased gradually and again the length of wire between the bridges noted for various weight to get readings of *l* for decreasing weight.
8. The diameter of the sonometer wire is measured at various points with the help of screw gauge and the average taken. The radius of the wire is half of the diameter.

OBERVATIONS

Mass of the ring = ... gm

Density of the material of the sonometer wire ρ = ... gm/cm^3

Table 4.1

S.No.	*Mass in gram (m)*	*Tension applied in dyne T = mg*	*Length (in cm) 'l'*		*Mean 'l'= (A + B)/2 (cm)*
			Weight increasing (A)	*Weight decreasing (B)*	
1.					
2.					
3.					
4.					
5.					
6.					

$$\text{Pitch} = \frac{\text{Distance travelled on the pitch scale}}{\text{Number of rotations of circular head}} = \text{..............} \text{ cm}$$

(If 2 rotations are given to the Circular scale, the distance moved is 1 mm, the Pitch = ½ = 0.5 mm.)

Number of divisions on the circular scale = ...

$$\text{Least count of screw gauge} = \frac{\text{Pitch}}{\text{Total number of divisions on circular scale}} = \text{............} \text{ cm}$$

Zero error of screw gauge =

I. If the zero of the C.S. lies below the base line of M.S. the error is positive.

II. If the zero of the C.S. lies above the base line of M.S. the error is negative.

Table 4.2. Measurement of diameter of sonometer wire.

S.No.	***Diameter along any direction 'a' (in cm)***			***Diameter along perpendicular direction 'b' (in cm)***			***Mean diameter d = (a + b)/2 (in cm)***	***Corrected diameter = d – (zero error with sign)***
	M.S.R.	***C.S.R.***	***TOTAL***	***M.S.R.***	***C.S.R.***	***TOTAL***		
1.								
2.								
3.								
4.								

Mean Diameter d = ... cm

Radius of wire r = ... cm

CALCULATIONS

Calculate the value of the frequency of the A.C. mains ν for each reading separately, by formula

$$\nu = \frac{1}{2l}\sqrt{\frac{Mg}{\pi r^2 \rho}}$$

1. ν_1 = ... Hertz
2. ν_2 = ... Hertz
3. ν_3 = ... Hertz
4. ν_4 = ... Hertz
5. ν_5 = ... Hertz
6. ν_6 = ... Hertz

Mean frequency ν = ... Hertz

RESULT

The frequency of A.C. mains as found by the

Experimental value ν = ... Hertz

Standard value ν = ... Hertz

$$\text{Percentage error} = \frac{\text{Standard value} \sim \text{Experimental value}}{\text{Standard value}} \times 100\%$$

PRECAUTIONS

1. There should be no kinks in the sonometer wire *i.e.,* the wire must be straight.
2. The pulley should move freely.
3. The magnet should be placed at the center of the wire without any contact with it.

CONCLUSION

VIVAVOCE

1. Why is the sound box provided with holes ?

2. What will happen if the pulley is not moving freely ?

3. What types of waves are set up in the sonometer wire ?

4. How will you test the resonance ?

5. What will happen if the bridges do not have sharp edges ?

6. What is node ?

7. What sources of A.C. frequency do you use ?

8. What do you mean by alternating current ?

9. What do you mean by the frequency of an alternating current ?
10. How will the frequency of A.C. mains get affected if a thicker wire is used with the sonometer ?
11. Is it true that the frequency of A.C. mains will be different if wire of different materials is taken with the sonometer ? Give answer with reasons.
12. Define wavelength, frequency, amplitude and the phase of a wave.
13. What are transverse and longitudinal waves ?
14. What do you mean by stationary wave ?
15. What are the laws governing the vibration in a stretched string ?
16. What do you mean by frequency of A.C. ?
17. What is the use of magnets in the experiment ?
18. Which law governs the force on the current carrying conductor in magnetic field ?
19. State and explain the Fleming's Left Hand rule.
20. Which material is use in the sonometer wire ?
21. Why cannot we use a ferromagnetic substance for wire ?
22. What is resonance ?
23. What is the principle involved in the experiment ?

❑❑❑

EXPERIMENT—05

AIM

To determine the value of *H*, horizontal component of earth's magnetic field by deflection magnetometer and vibration magnetometer.

APPARATUS REQUIRED

A deflection magnetometer, a vibration magnetometer, a bar magnet rectangular brass bar, a Vernier Callipers, a stopwatch or a stopclock, a balance, a weight box, a spirit level and a piece of chalk.

Fig. 5.1. *Vibration and deflection magnetometer.*

DESCRIPTION OF APPARATUS

Deflection Magnetometer

Deflection magnetometer consists of a small magnetic needle *NS* pivoted at the center of a circular graduated scale divided into four quadrants each reading from 0° to 90° so as to turn freely about a vertical axis. A long aluminum pointer, flat at the center and made pointed at its ends so as to read the deflection correctly and to provide the necessary damping is attached at right angles to the magnetic needle. The needle is taken small so as to rotate in a uniform field. It is enclosed in a box of non-magnetic material, which is fitted with a glass top so as to observe the scale and to avoid the needle from air currents. A plane mirror is fixed at the base of the compass box to read the position of the pointer normally. Thus, compass box is mounted on a wooden base with two long arms *A* and *B* on either side as shown in Fig. 5.2. Both the arms are fitted with a half meter scale graduated in cms, with their zero mark beginning from the center of the magnetic needle.

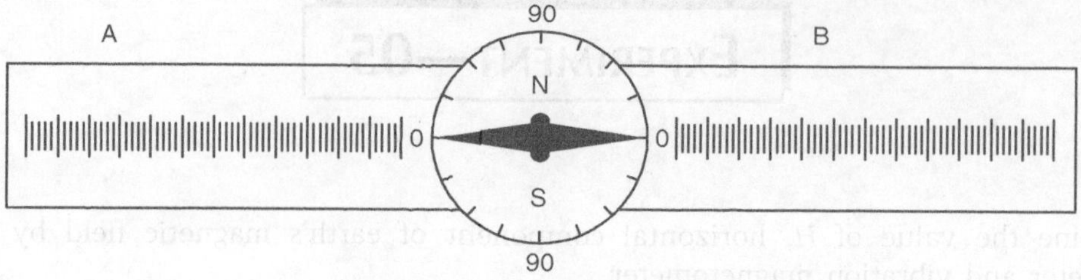

Fig. 5.2. *Deflection magnetometer.*

Vibration Magnetometer

It consists of a rectangular wooden box provided with levelling screws on the base, and having glass sides to prevent air currents from disturbing the magnet. A vertical and long brass or glass tube is fixed at the middle on the top of the box. It has got a torsion head at its upper end from which a metallic stirrup or pan is suspended to carry the magnet, and it is hung by silk fiber with torsion head. The magnet can be placed on stirrup by opening the glass window attached to the sides of the box (Fig. 5.3). At the base of the box a plane mirror strip etched with a straight line is also fitted. For setting the vibration of the magnet a long glass slit is also fitted at the centre of the top of the box along its length. When a bar magnet is suspended freely so as to oscillate in the horizontal plane and position of rest in magnetic meridian, pointing north of the magnet $T = 2\pi\sqrt{\frac{1}{MH}}$ gives time period of oscillation. As the time of oscillation of the magnet is given by $T = 2\pi\sqrt{\frac{1}{MH}}$.

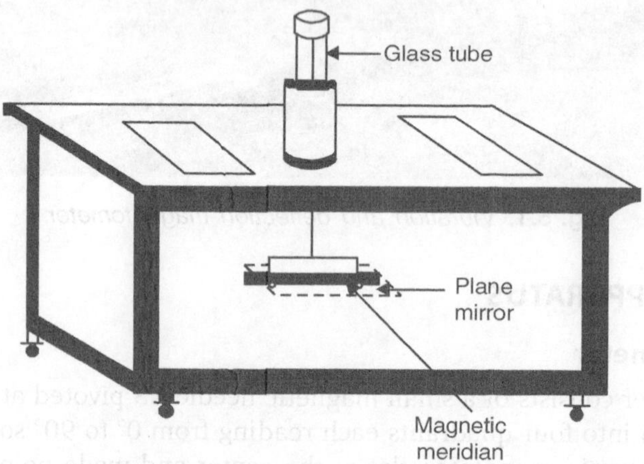

Fig. 5.3. *Vibration magnetometer.*

PRINCIPLE

A Magnetic needle is kept free to rotate; it rests in the *NS* direction in the absence of any other field due to the horizontal component of earth's magnetic field. If another magnetic field is applied in a direction perpendicular to the direction of horizontal component of the magnetic field then the needle is deflected and comes to the equilibrium position making an angle of θ with the horizontal component of earth's magnetic field and

$$F = H \tan \theta$$

Knowing the field acting due to the magnet the value of horizontal component of earth's magnetic field can be calculated.

THEORY

The field F due to a magnet of a magnetic moment M and length $2l$, kept in the EW direction (shown in Fig. 5.4), at a point along the axis of the magnet at a distance d from its center is given by the relation

$$F = \frac{2Md}{(d^2 - l^2)^2}$$

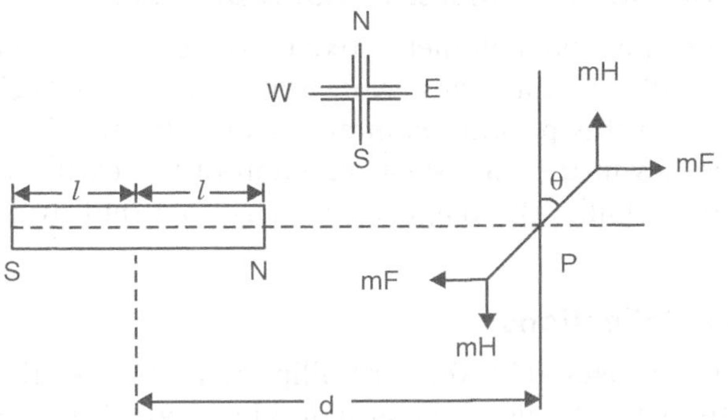

Fig. 5.4

The field F, due to magnet acts along WE and the field H due to earth acts along SN. Thus the needle kept at P is subjected to two fields at right angles to each other and will come in a position inclined at an angle θ, to the direction of H, when Tangent Law ($F = H \tan \theta$) is satisfied. Thus,

$$\frac{2Md}{(d^2 - l^2)^2} = H \tan \theta$$

$$\frac{M}{H} = \frac{(d^2 - l^2)^2 \tan \theta}{2d} \qquad ...(1)$$

Now, if the same magnet is allowed to oscillate freely, the time period T of oscillation is given by the relation

$$T = 2\pi\sqrt{\frac{1}{MH}}$$

or

$$MH = \frac{4\pi^2 I}{T^2} \qquad ...(2)$$

where I is the moment of inertia of the bar magnet having rectangular cross-section and is given by

$$I = \frac{w(a^2 + b^2)}{12}$$

In this relation, w is the mass of the magnet a is its length and b is the breadth of the magnet.

Dividing equation (2) by equation (1), we obtain

$$H^2 = 4\pi^2 \frac{I}{T^2} \frac{2d}{(d^2 - l^2)\tan\theta}$$

or

$$H = \frac{2\pi}{T(d^2 - l^2)}\sqrt{\frac{2Id}{\tan\theta}} \quad ...(3)$$

PROCEDURE

1. Method for setting the magnetometer in tan *A* position

Keeping arms of the deflection magnetometer fixed, rotate the compass box till (0–0) of the circular scale becomes parallel to the arms. Now rotate the arms such that the pointer reads (0–0) on the circular scale. In this position pointer is along the arms, magnetic needle is at right angles to the arms, which lies in the East West direction of the earth. Mark the boundary of the magnetometer by a piece of chalk. The magnetometer is now said to have been set up in tan *A* position.

2. For noting down the deflections

The length of the magnet is measured by Vernier Callipers. Then place the magnet on the east arm of the magnetometer with its length along the arm and its *N* pole [Fig. 5.5(*a*)] towards the needle at such a distance so as to obtained deflection nearly 45° (between 30° and 60°). Read both ends of the pointer by keeping the eyes normally such that the pointer and its image appear to be in the same plane. This avoids the error due to magnetic needle and pointer not being pivoted at the centre of the circular scale.

Now change the magnet pole as shown in Fig. 5.5(*b*), keeping the magnet on the same arm. Note the readings of both pointers again. This avoids the error due to wrong marking of the centre of the magnet and the asymmetrical position of poles.

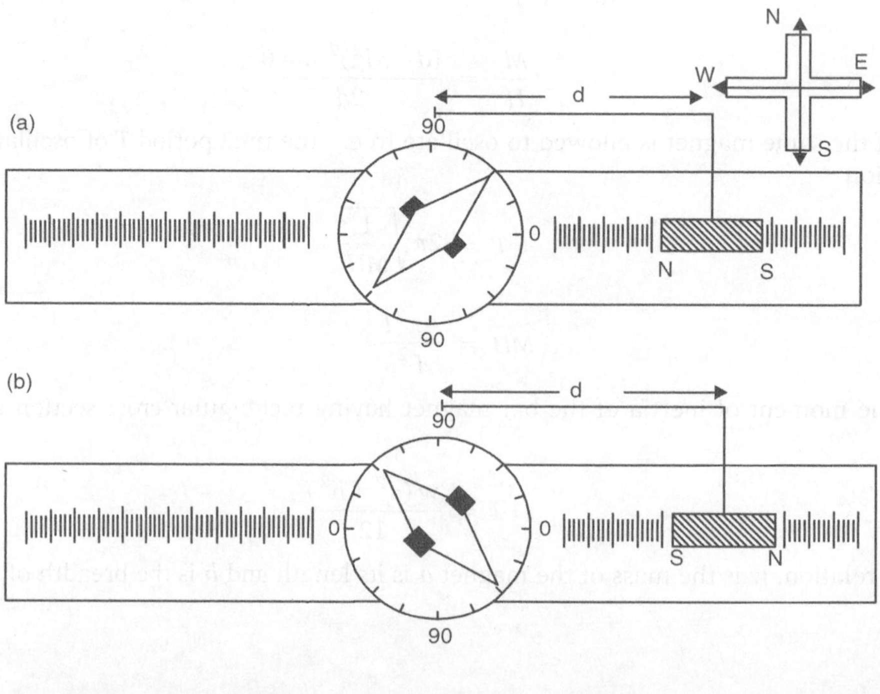

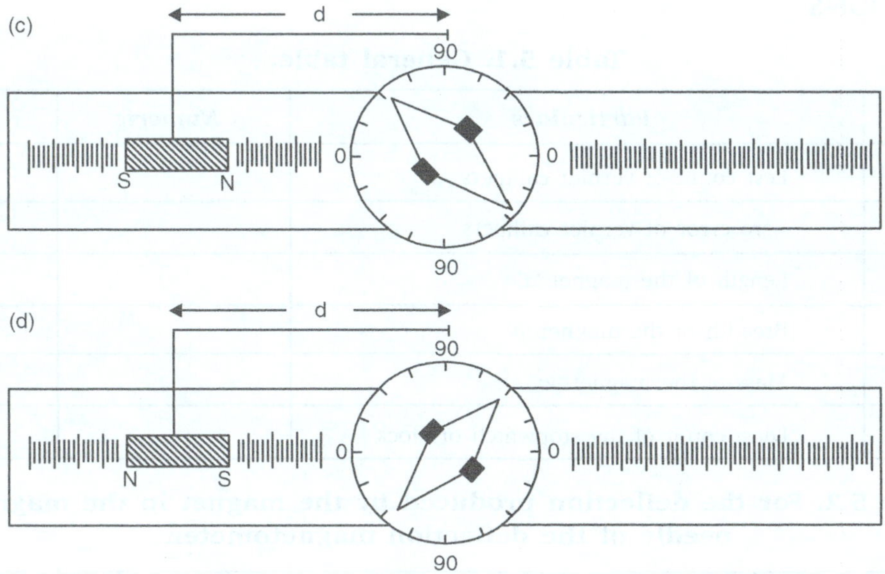

Fig. 5.5

Next, keep the magnet on the second arm [Fig. 5.5(*c*) and Fig. 5.5(*d*)] at the same distance and take four readings more, in the same way as described above. This eliminates the error due to non-coincidence of the centre of the circular scale with the zero of the linear scale. Mean of eight readings gives θ as in Table 5.2 of observations.

3. For finding time period of oscillation of bar magnet

1. First the vibration magnetometer is made horizontal by adjusting the levelling screws provided at its base with the help of spirit level.
2. Now the magnetometer is adjusted in magnetic meridian. For this place a compass needle on the reference line drawn on plane mirror at its base such that the centre of needle lays on the reference line. Then turn the apparatus in a horizontal plane so that the compass needle becomes exactly parallel to the reference line. Draw the boundary of the magnetometer with a chalk.
3. Then remove the twist in the suspension thread by keeping the brass rod on the stirrup and turning the knob so that the brass rod rests in direction parallel to the reference line. Now the rod is in magnetic meridian.
4. Now place the given magnet in the stirrup with its north pole facing towards geographic north.
5. Now displace this magnet slightly by bringing another magnet outside it, so that the suspended magnet starts oscillating. Note the time for 20 oscillations using a stopwatch three-four times.
6. Measure the mass *w* of the given magnet with the help of physical balance.
7. Use vernier callipers to find the length *a* and breadth *b* of the magnet.

OBSERVATIONS

Table 5.1. General table.

S.No.	*Particulars*	*Numeric*	*Unit*
1.	Lest count of vernier callipers		
2.	Zero error of vernier callipers		
3.	Length of the magnet '*a*'		
4.	Breadth of the magnet '*b*'		
5.	Mass of the magnet '*w*'		
6.	Least count of the stopwatch or clock		

Table 5.2. For the deflection produced by the magnet in the magnetic needle of the deflection magnetometer.

<table>
<tr><th rowspan="3">S.No.</th><th rowspan="3">Distance of mid-point of magnet d (in cm)</th><th colspan="4">Magnet on east arm</th><th colspan="4">Magnet on west arm</th><th rowspan="3">Mean deflection θ</th><th rowspan="3">tan θ</th></tr>
<tr><th colspan="2">N pole of magnet facing the compass box</th><th colspan="2">S pole of magnet facing the compass box</th><th colspan="2">N pole of magnet facing the compass box</th><th colspan="2">S pole of magnet facing the compass box</th></tr>
<tr><th>ϕ_1°</th><th>ϕ_2°</th><th>ϕ_3°</th><th>ϕ_4°</th><th>ϕ_5°</th><th>ϕ_6°</th><th>ϕ_7°</th><th>ϕ_8°</th></tr>
<tr><td>1.</td><td></td><td></td><td></td><td></td><td></td><td></td><td></td><td></td><td></td><td></td><td></td></tr>
<tr><td>2.</td><td></td><td></td><td></td><td></td><td></td><td></td><td></td><td></td><td></td><td></td><td></td></tr>
<tr><td>3.</td><td></td><td></td><td></td><td></td><td></td><td></td><td></td><td></td><td></td><td></td><td></td></tr>
</table>

Table 5.3. For the time period of magnet by the vibration magnetometer.

Least Count of stopwatch = ... sec

S.No.	*No. of oscillation n*	*Total time t (in sec)*	*Time period T = t/n (in sec)*
1.			
2.			
3.			

Mean T = ... sec

CALCULATIONS

1. First calculation are done for the moment of inertia of the bar magnet using the relation

$$I = \frac{w(a^2 + b^2)}{12}.$$

2. Secondly the value of H is calculated for the first reading of deflection θ and it's corresponding d, using the relation

$$H = \frac{2\pi}{T(d^2 - l^2)} \sqrt{\frac{2Id}{\tan\theta}}.$$

3. The same is done for all other readings of θ and corresponding d.
4. Mean H is calculated.

RESULT

The value of horizontal component of earth's magnetic field = ... gauss.

PRECAUTIONS AND SOURCES OF ERROR

1. The magnetometer needle should rotate freely and the arms should be steady.
2. The deflection magnetometer or vibration magnetometer once set should not be disturbed.
3. No magnet or magnetic material should be near the magnetometer during the experiment.
4. Magnet should always be kept on the arms of deflection magnetometer in the east west direction.
5. Compass box should be slightly tapped before taking any reading.
6. Eyes should be kept normally *i.e.,* the image of the pointer should be exactly below it.
7. Both ends of the pointer should be read.
8. Deflection should be nearly 45° and in no case it should be less than 30° and more than 60°.
9. Only half the length or (effective length of magnet 5/6th of Geometric length) should be substituted for l.
10. d should be measured from middle-point of the magnet.
11. Deflections when the magnet is changed to 2nd arm should be nearly same.

CONCLUSION

VIVAVOCE

1. What is the principal on which the deflection magnetometer works ?
2. How do you adjust the deflection magnetometer in tan A position and then how does tangent law hold ?
3. Why do you keep the deflection magnetometer only in tan A position in your experiment ?
4. Why is there a plane mirror underneath the magnetic needle in the deflection magnetometer ?
5. The magnetic needle in the compass box is very short while a pointer is long. Why ?
6. Why do you prefer to keep the deflection in the range of 30° to 60° ?
7. Can you use a wire for suspension in vibration magnetometer ?
8. Why is there a straight line drawn in the middle of base parallel to its length ?
9. On what factor does the time period of a magnet depends ?
10. Is the value of H same everywhere ?
11. What is meant by magnetic meridian ?
12. What is tangent law ?

❑❑❑

EXPERIMENT—06

AIM

To determine the moment of inertia of a flywheel about its own axis of rotation.

APPARATUS REQUIRED

Flywheel, slotted weights, thread, stopwatch, vernier callipers.

Fig. 6.1. *Flywheel setup.*

DESCRIPTION OF THE APPARATUS

The flywheel is shown in Fig. 6.1. It is basically a heavy wheel having a long axis and the mass distribution is such that the larger mass is concentrated at a relatively larger distance from the axis. The long axis is supported in bearings to keep the axis at a definite height from the ground and to keep it always horizontal. The mounting of the flywheel is such that the centre of gravity of the wheel is always on the axis of rotation and in this condition flywheel remains at rest in any position.

PRINCIPLE

The energy given to flywheel is used in providing it kinetic energy and overcoming the force of friction at the bearings. The kinetic energy of the flywheel is a function of its moment of inertia and angular velocity. If the energy provided to the flywheel and the angular velocity of the flywheel is known, the moment of inertia can be found.

THEORY

When the mass attached to the string falls due to gravity, it looses its potential energy. The loss of potential energy by the mass is proportional to the mass and the height it descends. This energy is used for the following three purposes :

1. To provide the kinetic energy to the descending mass.
2. To provide the kinetic energy to the flywheel.
3. To work against the friction.

Consider the translational velocity of the descending mass is v, the angular velocity of the flywheel is ω, the work done against the friction per tern is F (provided the work done against friction is steady), n_1 is the number of terns the flywheel rotates before the mass leaves the axis and h is the height traversed by the mass before leaving the axis than

The loss of potential energy $= mgh$

The kinetic energy gained by the mass $= \frac{1}{2}mv^2$

The kinetic energy of the flywheel $= \frac{1}{2}I\omega^2$

The work done against friction $= n_1F$

So according to the law of energy conservation

$$mgh = \frac{1}{2}mv^2 + \frac{1}{2}I\omega^2 + n_1F \quad \text{...(1)}$$

When the mass leaves the axel the only force acting against the motion of the flywheel is the force of friction. This force eventually brings the flywheel at rest after a certain time t, when the flywheel has completed n_2 rotations. So the work done by the force of friction must be equal to the kinetic energy gained by the flywheel, so

The work done by the friction is $= n_2F$

Hence,
$$n_2F = \frac{1}{2}I\omega^2$$

so
$$F = \frac{I\omega^2}{2n_2} \quad \text{...(2)}$$

Substituting the value of F in equation (1)

$$mgh = \frac{1}{2}mv^2 + \frac{1}{2}I\omega^2 + \frac{n_1 I\omega^2}{2n_2}$$

$$mgh - \frac{1}{2}mv^2 = \frac{1}{2}I\omega^2\left(1 + \frac{n_1}{n_2}\right)$$

$$2mgh - mv^2 = I\omega^2\left(1 + \frac{n_1}{n_2}\right)$$

$$I = \frac{(2mgh - mv^2)}{\omega^2 \left(1 + \frac{n_1}{n_2}\right)} \quad ...(3)$$

If r is the radius of the axel of the flywheel than the relation between the translational velocity of the descending mass and the rotational velocity of flywheel is

$$v = r\omega \Rightarrow v^2 = r^2\omega^2$$

Substituting the value in equation (3)

$$I = \frac{(2mgh - mr^2\omega^2)}{\omega^2 \left(1 + \frac{n_1}{n_2}\right)}$$

$$I = \frac{m\left(\frac{2gh}{\omega^2} - r^2\right)}{\left(1 + \frac{n_1}{n_2}\right)} \quad ...(4)$$

In equation (4) m, h, r, n_1 and n_2 can be obtained by the experiment and g is a constant. The average angular velocity can be determined by the following way.

After the mass leaves the axel the angular velocity starts decreasing from ω and ultimately becomes zero. So the average angular velocity is

$$\frac{(\omega + 0)}{2} = \frac{\omega}{2} \quad ...(5)$$

Meanwhile the flywheel completes n_2 rotations in time t so the angular distance travelled is $2\pi n_2$ so the angular velocity is

$$\frac{2\pi n_2}{t} \quad ...(6)$$

From equations (5) and (6)

$$\frac{\omega}{2} = \frac{2\pi n_2}{t}$$

$$\omega = \frac{4\pi n_2}{t} \quad ...(7)$$

FORMULA USED

Potential energy of a mass is converted into rotational kinetic energy of flywheel and translational kinetic energy of suspended mass so that

$$mgh = \frac{1}{2}mv^2 + \frac{1}{2}I\omega^2 + n_1 F$$

which reduces to

$$I = \frac{m\left(\frac{2gh}{\omega^2} - r^2\right)}{\left(1 + \frac{n_1}{n_2}\right)}$$

where I = Moment of inertia of given fly wheel about its own axis.

v = Linear velocity of mass suspended.

h = Distance moved by mass m before the string leaves the axle ($h = 2\pi r n_1$)

ω = Angular velocity = $4\pi n_2/t$

r = Radius of the wheel axle.

n_1 = Number of rotation made by the wheel till the mass leaves.

n_2 = Number of rotation made by the wheel after the mass had detached.

t = Time in sec for n_2 completing rotation.

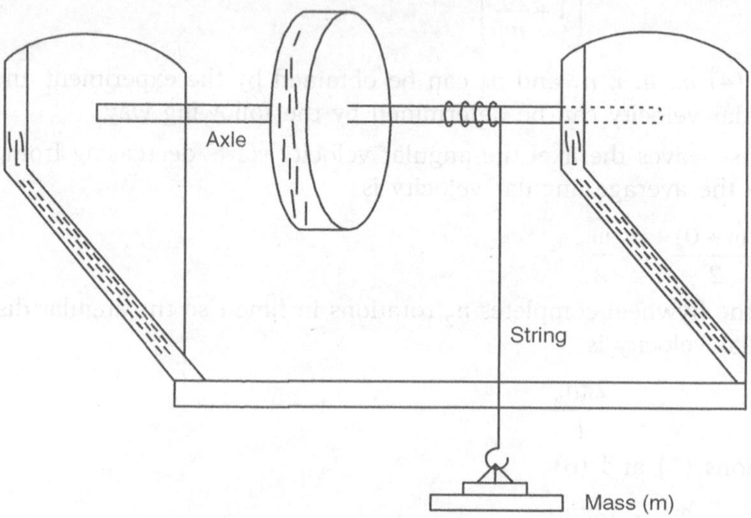

Fig. 6.2. *Experimental arrangement.*

PROCEDURE

1. First measure the radius of the axle using vernier callipers.
2. A string is then bound on the axle. One end of which is fixed on the axle and the mass is suspended on the other end.
3. Flywheel is held by hand to prevent rotation. Number of terns of the string (n_1) is counted.
4. If the flywheel is now released, it rotates as the mass moves downwards. It continues to rotate even after the mass leaves the axle. Number of these rotations (n_2) is counted.
5. Time period t for these rotation (n_2) are noted with the help of the stopwatch.

OBSERVATION

Vernier constant of the vernier scale $= \dfrac{\text{Value of smallest division of main scale}}{\text{Total number of vernier scale divisions}}$.

Table 6.1. For the radius of the axle.

S.No.	*Reading of diameter along any direction*			*Reading of diameter along perpendicular direction*			*Mean diameter (a + b)/2*	*Mean radius*
	M.S.R.	*V.S.R.*	*TOTAL (a)*	*M.S.R.*	*V.S.R.*	*TOTAL (b)*	*(cm)*	*(cm)*

Table 6.2. For n_1 and n_2.

S.No.	*Mass (in gram)*	n_1	n_2	$h = 2\pi r n_1$ *(in cm)*	*Time for* n_2 *(t) (in second)*

CALCULATIONS

Substituting values of various parameters in the formula :

$$I = \frac{m\left(\dfrac{2gh}{\omega^2} - r^2\right)}{\left(1 + \dfrac{n_1}{n_2}\right)}.$$

RESULT

Moment of inertia of the given flywheel about it's own axis of rotation is found to be ......gm/cm^2.

PRECAUTIONS

1. Extra care must be taken to start or stop the stopwatch.
2. Fractional value of rotation should be avoided by adjusting the vertical height.
3. The friction at the bearing should be avoided.
4. The loop slipped over the peg should be quite loose.

CONCLUSION

VIVAVOCE

1. How can you define moment of inertia ?
2. What is the physical significance of moment of inertia ?
3. What is a flywheel ?
4. What do you mean by radius of gyration ?
5. What is the unit of moment of inertia and radius of gyration ?
6. What precaution do you take in winding the thread on the axel ?
7. What happens if the turns of the thread are winded one over the other ?
8. How moment of inertia of a body depends on the mass and radius of gyration ?
9. State parallel axis theorem.
10. State perpendicular axis theorem.
11. About which axis the moment of inertia of the flywheel is minimum ?
12. What are the assumptions made and how can you justify them ?

❑❑❑

EXPERIMENT—07

AIM

To accurately measure the acceleration due to gravity g with the help of a reversible physical pendulum (or bar pendulum).

APPARATUS REQUIRED

Bar pendulum, stopwatch, knife-edges fixed to a rigid support, traveling microscope and meter scale.

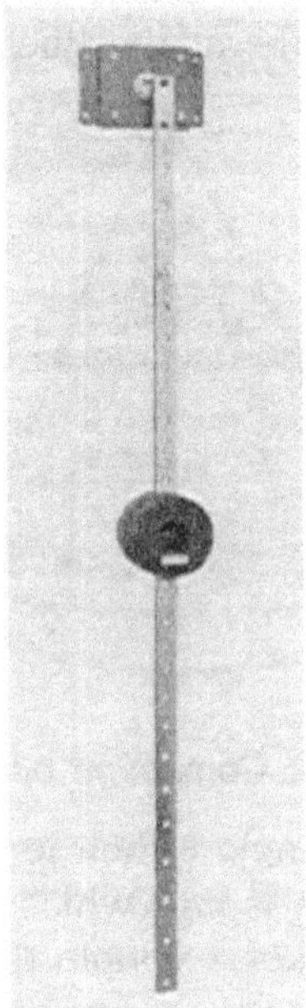

Fig. 7.1. *Bar pendulum.*

DESCRIPTION OF APPARATUS

A bar pendulum is shown in Fig. 7.1. It consists of a uniform rectangular iron or aluminum heavy bar about one meter long with a large number of holes drilled along its length at equal distances 5-5 cm from each other. The holes are symmetric about center of gravity G. A sharp knife-edge is attached to some heavy frame provided with leveling screws to make the knife-edge horizontal. The bar can be suspended with a horizontal knife-edge in one of its hole such that it can oscillate freely. Such a hole is called center of suspension. The time period about each hole on one end of G (center of gravity) and at the same time the time period about holes on the other side at equal distance from G (center of gravity) can be determined by hanging the bar pendulum at different holes. The time period of oscillation is measured by stopwatch. A traveling microscope can be used for counting number of oscillations (because angle of displacement is very small).

PRINCIPLE

The principle is based on the interchangeability of the centers of suspension and center of oscillation. We know that for a point of suspension, there is another point on the other side of center of gravity, known as center of oscillation about which the time period is the same. The distance between center of suspension and center of oscillations is known as the length of equivalent simple pendulum $L = l + \left(\frac{K^2}{l}\right)$. Knowing this distance g can be calculated.

THEORY

Consider a rigid body which can rotate freely about a horizontal axis through point S. Let m be the mass of the body and l is the distance of center of gravity G from the point of suspension S. At rest position, center of gravity G will be at the line SN.

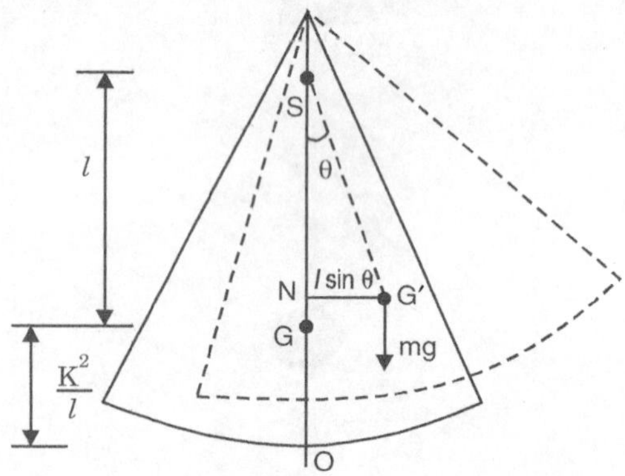

Fig. 7.2. *Compound pendulum.*

When the body is displaced by an angle θ then its center of gravity shifted from G to G'. In this position the weight of the pendulum is mg (which is directed downward at G') and a force of reaction on point of suspension S makes a couple that is called restoring couple (torque).

$$\begin{aligned} \therefore \quad \text{Torque} &= \text{Force} \times \text{Perpendicular distance} \\ &= mg \times NG' \\ &= mg \times l \sin\theta \\ &= mgl\,\theta \text{ (because } \theta \text{ is very small so } \sin\theta \approx \theta) \qquad \text{...(1)} \end{aligned}$$

This torque tends to force pendulum back to its mean position due to which angular acceleration $\alpha = \frac{d^2\theta}{dt^2}$ produced in the body. If I is the moment of inertia of the body about an axis through S perpendicular to the plane of paper, then

$$\text{Restoring couple} = I\frac{d^2\theta}{dt^2}$$

In equilibrium position,

$$I\frac{d^2\theta}{dt^2} = -mgl\,\theta$$

or $$\frac{d^2\theta}{dt^2} + \frac{mgl}{I}\theta = 0$$

or $$\frac{d^2\theta}{dt^2} + \omega^2\theta = 0 \qquad \text{...(2)}$$

where, $$\omega^2 = \frac{mgl}{I}$$

Hence the body executes simple harmonic motion and its time period is given by

$$T = \frac{2\pi}{\omega} = 2\pi\sqrt{\frac{I}{mgl}} \qquad \text{...(3)}$$

If $I_g = mK^2$ is the moment of inertia of the body about a parallel axis through center of gravity G, then by theorem of parallel axes the moment of inertia I of the body about vertical axis passing through point of suspension S is given by

$$\begin{aligned} I &= I_g + ml^2 \\ &= mK^2 + ml^2 \\ &= m(K^2 + l^2) \end{aligned}$$

where K is the radius of gyration about a parallel axis passing through the center of gravity G and l is the distance between point of suspension and center of gravity.

Substituting the value of I in equation (3), we have

$$\begin{aligned} T &= 2\pi\sqrt{\frac{m(K^2 + l^2)}{mgl}} \\ &= 2\pi\sqrt{\frac{(K^2 + l^2)}{gl}} \\ &= 2\pi\sqrt{\frac{(K^2/l) + l}{g}} \qquad \text{...(4)} \end{aligned}$$

Obviously, this time period is equivalent to the time period of that simple pendulum of length $L = (K^2/l) + l$, therefore the length $L = (K^2/l) + l$ is called the length of equivalent simple pendulum and time period may be written as,

$$T = 2\pi\sqrt{\frac{L}{g}} \qquad \text{...(5)}$$

If whole mass of the compound pendulum is supposed to be concentrated at point O, at a distance $= l + (K^2/l)$, from the center of suspension S, it will behave like as a simple pendulum and will have same time period. The point O is called center of oscillation.

If now the compound pendulum is reversed and suspended at point O (*i.e.*, the point of suspension is changed from S to O, then the time period about this point is

$$T' = 2\pi\sqrt{\frac{\left(K^2/l'\right) + l'}{g}}$$

where, $OG' = l' = K^2/l$

$\therefore \quad K^2/l' = l$

$$\therefore \quad T' = 2\pi\sqrt{\frac{l + (K^2/l)}{g}} \quad \text{...(6)}$$

From equations (4) and (6) it is clear that

$$T = T'$$

i.e., the time period about the point of suspension and point of oscillation are equal. It means both the points are interchangeable.

By squaring equation (5), we get

$$T^2 = 4\pi^2 \frac{L}{g}$$

or,
$$g = 4\pi^2 \frac{L}{T^2} \quad \text{...(7)}$$

Thus, g can be determined from above expression.

FORMULA USED

The value of g can be calculated with the help of the following formula :

$$g = \frac{4\pi^2 L}{T^2}$$

where, $L = l + \left(\frac{K^2}{l}\right)$ = Distance between centers of oscillation and point of suspension

= The equivalent length of simple pendulum (it can be obtained from graph)

and T = Time period of the bar pendulum.

PROCEDURE

1. Hang the bar pendulum vertically by means of a knife-edge.
2. Allow the bar to oscillate through a small angle 3 to 4° with knife-edge passing through hole No. 1 and note the time for 20 oscillations with the help of a stopwatch and travelling microscope, and find the time period of one oscillation.
3. Similarly, find the time periods for other holes, till the center of gravity is reached.
4. Now, invert the bar pendulum and repeat the same procedure and find the time periods for the holes in this side.
5. Now, plot a graph between time period T and the distance of the holes from one end of the rod as shown in Fig. 7.3.

OBSERVATIONS

Table 7.1. For time period of the bar pendulum.

S.No.	Hole number	Distance of the hole from center of gravity (cm)	Number of oscillations	Time period T (Second)

GRAPHICAL CALCULATIONS

Plot a graph between time period T and the distance of that particular hole about which time period is measured as shown in Fig. 7.3. The distance of the hole is measured from one end of the bar, by taking origin at the center of the graph.

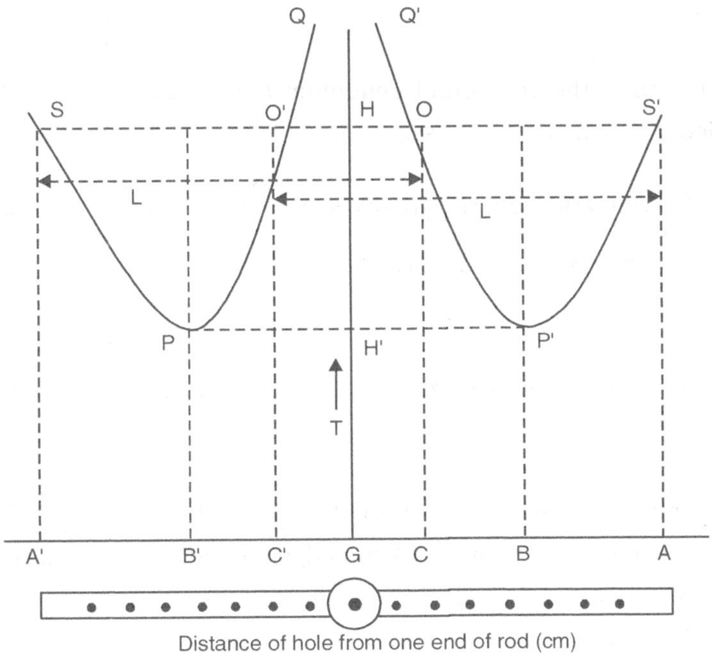

Fig. 7.3. *Graphical analysis.*

It is clear from the graph that :

(*i*) Two curves are QPS and $Q'P'S'$ are symmetric about the axis passing through the center of gravity G.

(*ii*) Initially as the distance of the hole from one end of the bar increases, the time period decreases and a point reaches where the time period becomes minimum. After this on increasing the distance of hole again, the time period increases and at center of gravity *G*, time period become maximum. Repeating the same process for other side of *G*, initially time period decreases and reaches to minimum value and then increases again. Thus, we have two identical curve *QPS* and *Q′P′S′*.

(*iii*) If a horizontal line is drawn parallel to *X*-axis, it intersects the two curves at four points *S*, *O′*, *O* and *S′* about which the time periods are the same. The distance between *S* and *O* is *L*, as is the distance between *S′* and *O′*. If the point *S* is center of suspension then *O* is the center of oscillation and the distance between the two is the length of the equivalent simple pendulum *L* corresponding to these points, time periods may be obtained from period axis.

(*iv*) Effective length of equivalent simple pendulum is $L = S'O' = SO$.

(*v*) Now *g* can be calculated by applying the formula;

$$g = \frac{4\pi^2 L}{T^2}$$

(*vi*) If the lowest points of both the curves are joined, then $PP' = 2K$ and $T_{min} = GH'$.

CALCULATIONS

From the graph,

The effective length of the compound pendulum L = ... cm

The time period T = ... sec

$\therefore$ The value of acceleration due to gravity $g = \dfrac{4\pi^2 L}{T^2}$ = ... cm/sec^2

and the radius of gyration K = ... cm

RESULT

The value of acceleration due to gravity *g* at .... = ... cm/sec^2

PRECAUTIONS

1. The support must be massive and absolutely unmoving.
2. Before starting the experiment, the knife-edge should be made horizontal.
3. The knife-edge must not "wander" in position on its support.
4. The material of the bar must be uniform throughout the length.
5. Use only small amplitude swings, less than 5°.
6. The amplitude of oscillations should be kept small.
7. The time of oscillations should be counted at least for 20 oscillations.
8. Curves on the graph should be drawn smoothly.

CONCLUSION

VIVAVOCE

1. What do you understand by gravity ?
2. What do you mean by acceleration due to gravity ? What is the unit of g ?
3. What is Newton's law of gravitation ?
4. What is moment of inertia and radius of gyration ?
5. What is parallel and perpendicular axis theorem ?
6. Why does the value of g vary at different places and at different distances from the surface of the earth ?
7. How does the value of g vary at different places and at different distances from the surface of the earth ?
8. What is a simple pendulum ?
9. Suppose a clear hole is bored through the center of the earth and a ball is dropped in it, what will happen to the ball ?
10. Define a compound pendulum. Which method is accurate for calculating g, simple pendulum or compound pendulum ?
11. What do you understand by center of suspension and oscillation ?
12. How does the period vary with distances of knife-edge from center of gravity of pendulum ?
13. What do you mean by equivalent simple pendulum ?
14. What is the length of equivalent simple pendulum ?
15. About how many and which points is the period of a compound pendulum the same ?
16. What is periodic time, when center of suspension coincide with center of gravity of compound pendulum ?
17. Why the amplitude of oscillations should be kept small ?

❑❑❑

SURFACE TENSION

INTRODUCTION

Surface tension arises because each molecule in a liquid is attracted to its fellow molecule. A molecule of the surface is attracted into the bulk liquid, which tends to reduce the liquid's surface area, just as a stretched membrane does (Fig. 1).

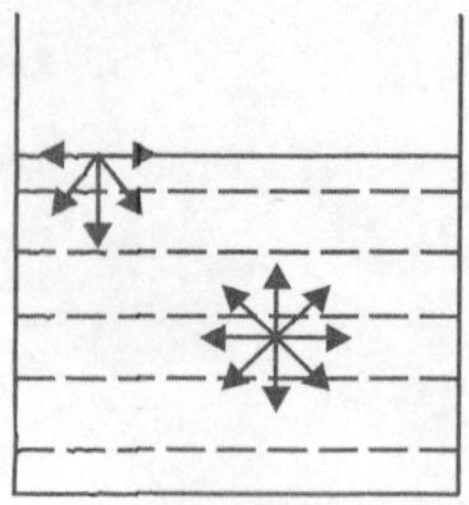

Fig. 1

To wash clothing thoroughly, water must be forced through the tiny spaces between the fibers. The required water pressure can be reduced by increasing the temperature of the water and adding soap, both of which decrease the surface tension (Fig. 2).

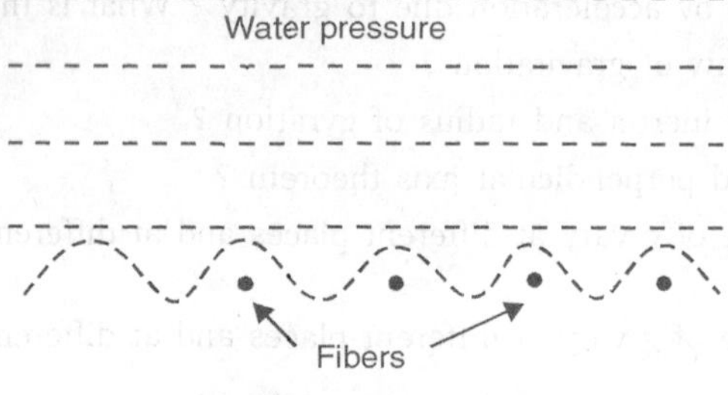

Fig. 2

IMPORTANCE

Surface tension is important for a millimeter sized water drop. The ratio of surface area to volume is $3/r$, which increases with decreasing radius (a spherical drop of radius r has surface area $4\pi r^2$ and volume $\frac{4\pi r^3}{3}$). For large quantities of liquid, however, this ratio is relatively small; hence surface tension is negligible compared to pressure forces.

DEFINITION

Surface tension can be defined as force acting per unit length of the line drawn on the surface of the liquid, the direction being at right angles to this line and tangential to the surface of the liquid. Hence *SI* unit is N/m (Fig. 3).

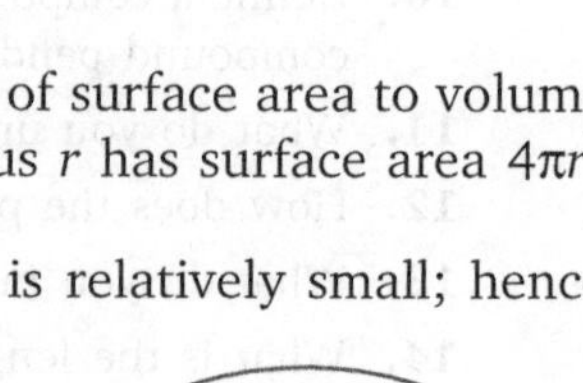

Fig. 3

When liquid comes in contact with a solid or some other liquid, its surface near the plane of contact in general becomes curved (Fig. 4). The contact angle can have any value between 0 and 180°.

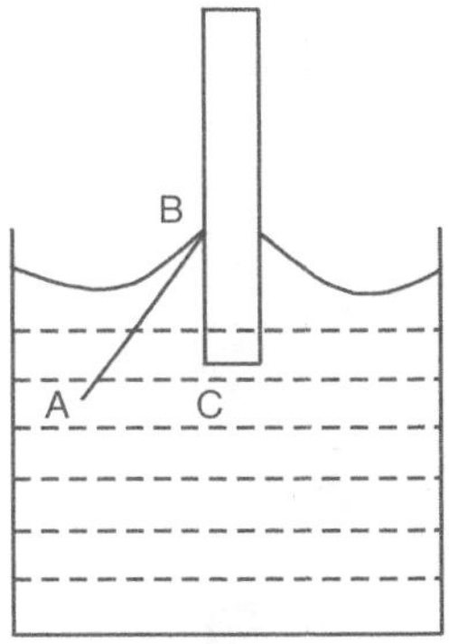

(a) $\angle ABC$ is acute

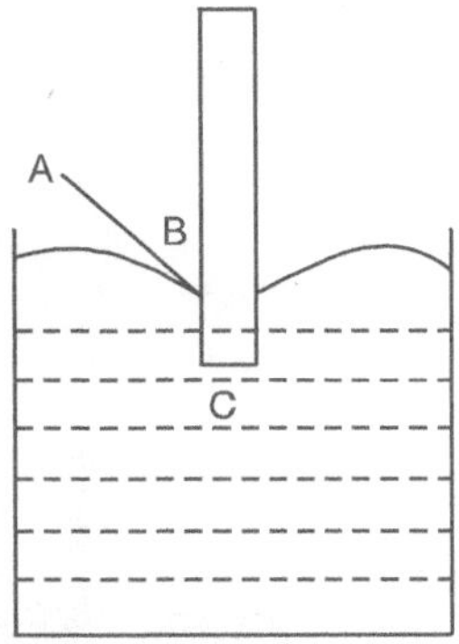

(b) $\angle ABC$ is obtuse

Fig. 4

As shown in Fig. 5(*a*) resultant force due to surface tension on a molecule at the surface is zero. Hence the cohesion pressure is just nominal. Whereas in Fig. 5(*b*) cohesion pressure is decreased as the resultant force is upward. For the convex force surface as in Fig. 5(*c*), the resultant force on the molecule at the surface of the liquid is downward and hence cohesion pressure is increased here.

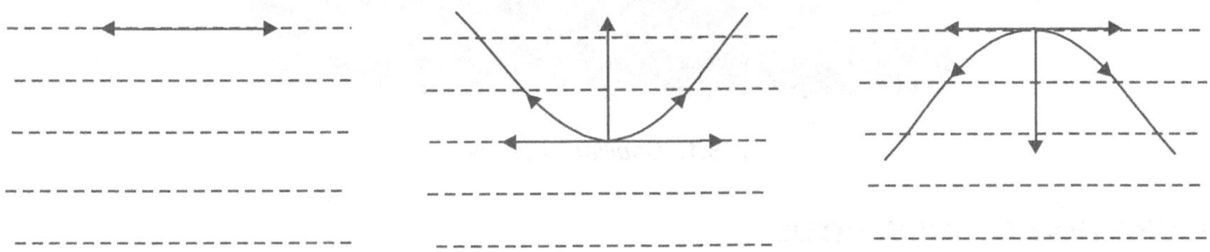

(a) (b) (c)

Fig. 5

EXPERIMENT—08

AIM

To determine the surface tension of water by capillary rise method.

APPARATUS REQUIRED

Capillary tube, beaker, water, travelling microscope, needle, rubber band, lamp and vertical stand.

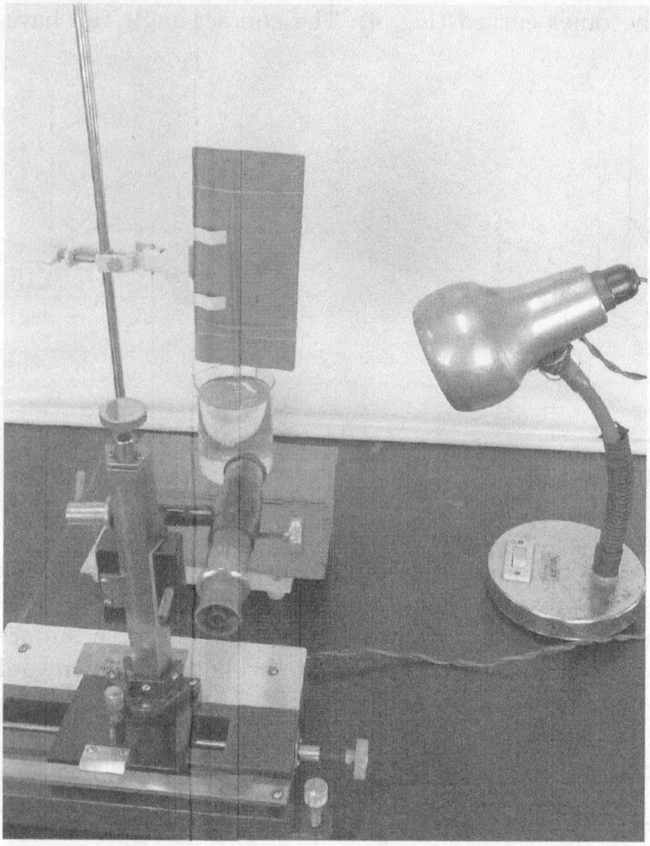

Fig. 8.1. *Capillary rise setup.*

DESCRIPTION OF APPARATUS

A capillary tube of uniform bore and a needle are passed through a piece of cork clamped to a retort stand and held vertically. One end of the capillary tube is immersed in fresh water contained in a beaker. The needle is kept parallel to the capillary tube and the lower end of the needle just makes contact with the free surface of water in the beaker. Lamp is used to illuminate the capillary tube (or object which is to be focused with travelling microscope). Microscope can be travelled along vertical and horizontal scale.

PRINCIPLE

When one end of a glass capillary tube of uniform diameter, is dipped into water, the water level in the tube rises to a certain height above the liquid level in the beaker. This phenomenon is also known as capillarity.

THEORY

Let r be the radius of the cylindrical capillary tube, d is density of water, T surface tension of water and θ the angle of contact of water-glass. The force due to surface tension is inclined with an angle θ with the wall of the tube and its direction is towards the liquid.

An equal reaction force acts on the water meniscus in upward direction. The action force (surface tension) is unable to pull the tube downwards, but the reaction remains unbalanced and hence pulls the water up. The water therefore goes on rising in the tube till the upward reaction is balanced by the weight of the liquid column raised.

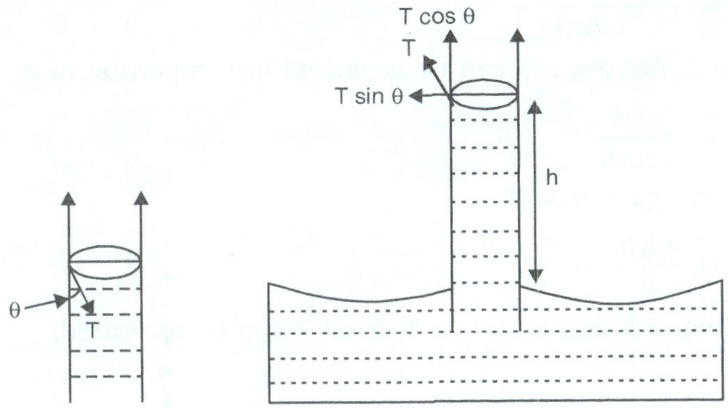

Fig. 8.2

Total upward force $= T \cos \theta . 2\pi r$ ...(1)

Volume of the water in the cylindrical column of height h and radius $r = \pi r^2 h$

As the tube is narrow, the radius of curvature of the meniscus is approximately equal to the radius r of the capillary tube (Fig. 8.3).

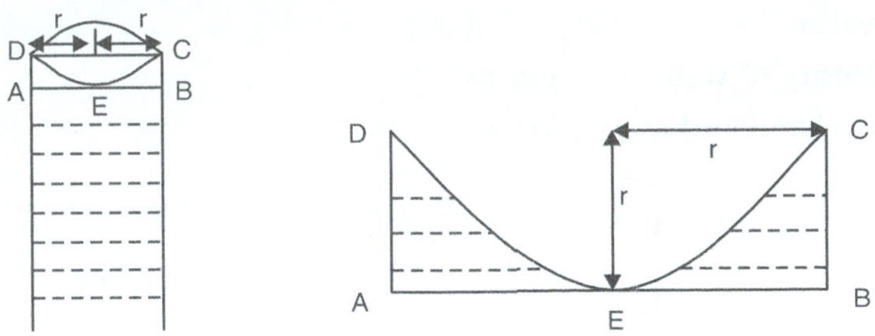

Fig. 8.3

Therefore, volume of the liquid under the meniscus as in Fig. 8.3 $= \left[\pi r^2 . r - \frac{1}{2}\left(\frac{4}{3}\pi r^3\right)\right] = \frac{1}{3}\pi r^3$

Therefore, total volume of the water in the capillary tube

$$= \pi r^2 h + \frac{1}{3}\pi r^3$$

$$= \pi r^2 (h + r/3)$$

Now, mass of the water raised in the tube (m = volume × density)

$$= \pi r^2 (h + r/3)d$$

Therefore, weight of the water in the tube

$$= mg = \pi r^2 (h + r/3)dg \qquad ...(2)$$

where, g is the gravitational acceleration.

In equilibrium, as the liquid is in stationary position, from equations (1) and (2)

Total upward force = Total downward force (weight of water raised in the tube)

$$T \cos \theta . 2\pi r = \pi r^2 (h + r/3) dg \quad \ldots(3)$$

or
$$T = \frac{rdg(h + r/3)}{2\cos\theta}$$

As the tube is very narrow, $r/3$ can be neglected in comparison to h

Therefore, $T = \dfrac{rdgh}{2\cos\theta}$

For pure water and glass $\theta = 0°$

Hence, $$T = \frac{rdgh}{2} \quad \ldots(4)$$

Knowing the value of h and r, surface tension T can be measured.

FORMULA USED

Surface tension of the water, $T = \dfrac{rdgh}{2}$

where,
- T = Surface tension of water at the room temperature.
- r = Inner radius of the capillary tube.
- h = Height of water column raised in capillary tube, measured vertically from the bottom of meniscus to the horizontal part of the free surface of the water in the beaker.
- d = Density of water = 1 gm/cc.
- g = Acceleration due to gravity.

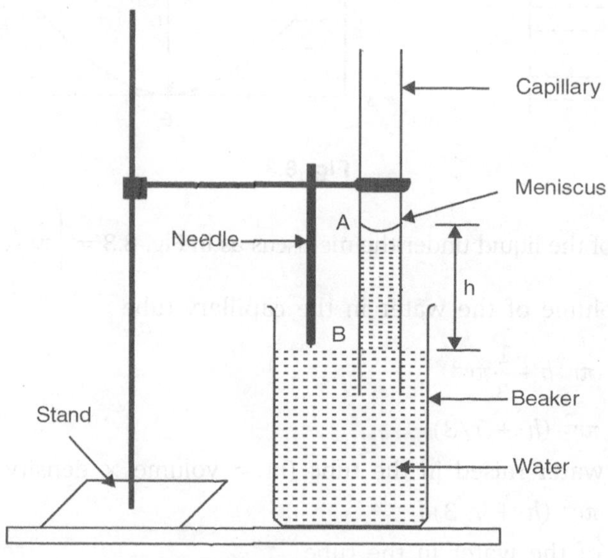

Fig. 8.4. *Experimental arrangement.*

PROCEDURE

1. Firstly clean capillary tube and beaker with nitric acid and dry in air.
2. Make arrangements as shown in Fig. 8.4.
3. Calculate least count of travelling microscope.
4. Focus cross-wire of travelling microscope on meniscus (upper level) of liquid column and take readings of vertical scale of travelling microscope.
5. Now, by moving microscope downwards and shifting horizontally, focus cross-wire on the tip of needle and take readings of vertical scale of travelling microscope in this position.
6. Now, remove capillary from the stand and clamp it horizontally so that its cross-section should be towards you as shown in Fig. 8.5.
7. Then, adjust travelling microscope in front of capillary so that a clear image of cross-section of the capillary tube must be visualize.
8. Now, focus cross-wire of travelling microscope on inner circumference in first position, then in second position as shown in Fig. 8.5. Take readings of horizontal scale of travelling microscope in both positions.
9. Measure temperature of the water in beaker with thermometer.

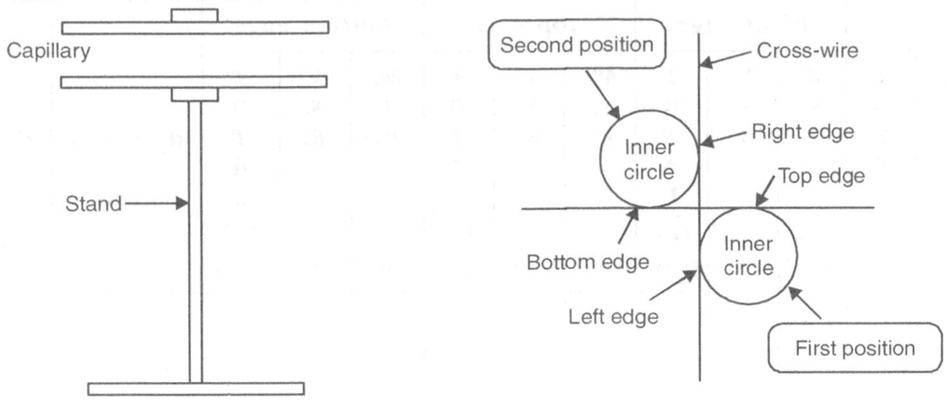

Fig. 8.5. *Capillary diameter measurement.*

OBSERVATION

- Temperature of water, t = ... °C
- Least count of the travelling microscope

  Value of 1 main scale division on main scale of travelling microscope, x = ... cm

  Total number of divisions on vernier scale, n = ...

$$\text{Least count of travelling microscope} = \frac{\text{Value of 1 main scale division}}{\text{Total number of divisions on vernier scale}}$$

$$= \frac{x}{n} \text{ cm} = \ldots\ldots \text{ cm}$$

Table 8.1. Determination of *h*.

S.No.	Microscope focused on						Height of the water level rise in the capillary $h = A - B$ (cm)
	Meniscus (A)			Tip of the needle (B)			
	M.S.R. (cm)	V.S.R.= V.S.D. × L.C.	Total Reading (cm)	M.S.R. (cm)	V.S.R.= V.S.D. × L.C.	Total Reading (cm)	
1.							
2.							
3.							

Mean h = ... cm

Table 8.2. Measurement of radius of capillary tube.

Microscope focused on												Diameter	
Left edge			Right edge			Top edge			Bottom edge				
M.S.R. (cm)	V.S.R. (cm)	TOTAL (a) (cm)	M.S.R. (cm)	V.S.R. (cm)	TOTAL (b) (cm)	M.S.R. (cm)	V.S.R. (cm)	TOTAL (c) (cm)	M.S.R. (cm)	V.S.R. (cm)	TOTAL (d) (cm)	$d_1 = a \sim b$ (cm)	$d_2 = c \sim d$ (cm)

Mean diameter, $d = \dfrac{(d_1 + d_2)}{2}$ = ... cm

Radius, $r = \dfrac{d}{2}$ = ... cm

CALCULATION

Surface tension of water, $T = \dfrac{rdgh}{2}$ = ...dyne/cm

RESULT

The surface tension of water at ...... °C is T = ... dynes/cm

PRECAUTIONS

1. The capillary tube should be of uniform bore.
2. The capillary tube and beaker should be clean and should not be contaminated with oil or grease.

3. The capillary tube must be kept vertical.
4. The top of the capillary tube must be open.
5. As the surface tension of a liquid varies with temperature, the temperature of water must be recorded.

CONCLUSION

VIVAVOCE

1. What do you understand by the phenomenon of surface tension ?
2. What is the cause of surface tension and what is its unit ?
3. What is the shape of the liquid surface when it is in contact with a solid ? When will the meniscus be convex or concave ? And why ?
4. What is angle of contact ? What are its units ? What is the value of angle of contact in the case of water and glass, mercury and glass ?
5. Why does water rise in a capillary tube and not in a water tube ?
6. Does every liquid rise in glass tube ? If not why ? How the rise will be affected by (*a*) change in the diameter in bore, (*b*) change in the shape of the bore and (*c*) the bore becoming non-uniform ?
7. From where, energy come when the liquid rises against gravity in a capillary ?
8. What are the advantages of this method over others ?
9. What do you understand by surface energy ?
10. Which molecules have higher surface energy ?
11. Is surface tension of a liquid constant ?
12. How is surface tension of a liquid affected by change in temperature ?
13. Water wets the fingers while mercury does not. Give reason.
14. What is capillary ? Does each liquid rise in a capillary ?
15. State factors on which rise of liquid in a capillary depends.
16. How is the height of water column in capillary affected if diameter of capillary is doubled ?
17. Give some examples of capillarity action.
18. Give some examples, which demonstrate the use of surface tension of liquids ?

❑❑❑

EXPERIMENT—09

AIM

To determine the surface tension of water at laboratory temperature by Jaeger's method.

APPARATUS REQUIRED

Jaeger's apparatus, capillary tube, a beaker with water, an adjustable stand and traveling microscope.

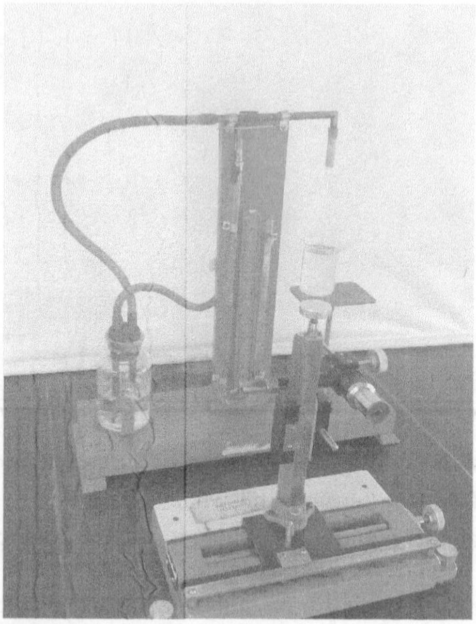

Fig. 9.1. *Jaeger's apparatus with traveling microscope.*

DESCRIPTION OF APPARATUS

The apparatus used for this experiment is shown in Fig. 9.1. It consists of a Woulff bottle which is a big glass bottle fitted with a cork on its mouth. The cork has two holes. Dropping funnel containing water is fitted in one hole of the cork. Through the other hole, a glass tube is attached. A manometer is fitted with as shown in Fig. 9.1. The end of the tube is drawn into a fine capillary. This capillary is dipped to a depth of 4-5 cm in the liquid in a beaker, whose surface tension is to be found out. The Woulff bottle is filled up to nearly $\frac{2}{3}$rd part of it with water by means of the funnel. The arms of the manometer are generally filled with experimental liquid (water).

PRINCIPLE

As the capillary tube dips in the liquid, the liquid rises in it and the shape of the meniscus is hemispherical. If the stop cork of the dropping funnel is slightly opened, the water slowly falls into the bottle. Consequently pressure increases in the bottle and the difference between the liquid levels in the arms of the manometer increases. Due to this, the liquid moves down in the capillary and forms a bubble at the end. These resultant forces in the curved surface would try to make them plane. Since these are curved in equilibrium, an excess pressure p is supposed to act opposite the resultant force T due to surface tension. The excess of pressure always acts towards the concave side of the curved surface as shown in Fig. 9.2.

THEORY

For a liquid drop, pressure inside it must be greater than the pressure outside it. If the excess pressure inside the liquid drop is p, r is the radius of the liquid drop and T is the surface tension, the upward thrust on the plane face $PQRS$ will be equal to

$$p\pi r^2 \text{ [Force = Pressure} \times \text{Area]} \qquad \text{...(1)}$$

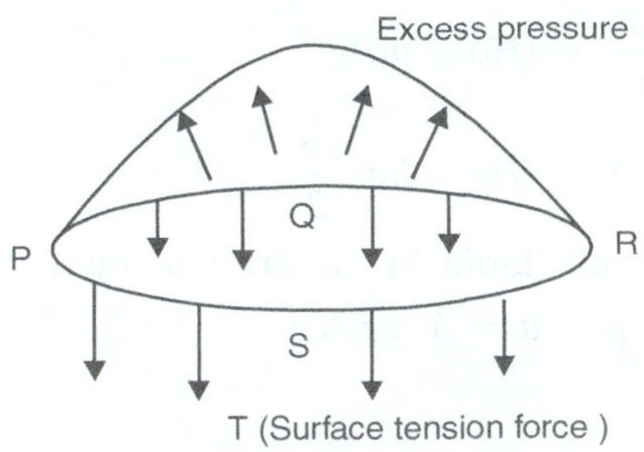

Fig. 9.2

For the equilibrium of upper half of the drop

$$p\pi r^2 = T2\pi r \qquad \text{...(2)}$$

where T is the force due to surface tension round its edge and acting downwards

From equation (2)

$$p = \frac{2T}{r} \qquad \text{...(3)}$$

In case of a bubble it is $p = \dfrac{4T}{r}$ ...(4)

[Because bubble is like a hollow spherical shell and each half of the bubble is in equilibrium. Hence in this case the force due to surface tension will be due to both these surfaces *i.e.*, it will be $2 \times 2\pi rT$].

From equations (3) and (4) we can conclude that smaller is the drop or bubble, more is the excess pressure inside it. In Jaeger's apparatus the pressure inside the bubble is measured with the help of the difference in manometer level and rises in maximum and then falls as the bubble break.

The radius of the bubble gradually decreases with increasing the pressure inside it, till it reaches the minimum value when the shape of the bubble is hemispherical of radius r equal to the radius of the opening at A (Fig. 9.3). The difference in the level of the liquid in the two limbs of the manometer is now maximum say H. The bubble is unstable because a slight increase in the radius decreases the internal pressure to produce equilibrium and as the external pressure is constant the bubble breaks away.

Just before the bubble breaks away, the pressure inside the bubble

$$p_1 = p_A + H\rho g \quad ...(5)$$

where p_A is the atmospheric pressure and ρ the density of the liquid in manometer.

At the same time pressure outside the bubble

$$p_2 = p_A + hdg \quad ...(6)$$

where h is the depth of capillary tube in the beaker and d the density of the experimental liquid.

Therefore, excess pressure inside the bubble from equations (5) and (6)

$$p = p_1 - p_2$$

$$= (H\rho - hd)g \quad ...(7)$$

In equilibrium, from equations (3) and (7)

$$\frac{2T}{r} = (H\rho - hd)g$$

or

$$T = (H\rho - hd)\frac{rg}{2} \quad ...(8)$$

If the experimental liquid and liquid in the arms of manometer is water, then

$$\rho = d = 1 \text{ gm/cc}$$

Then, from equation (8)

$$T = (H - h)\frac{rg}{2} \quad ...(9)$$

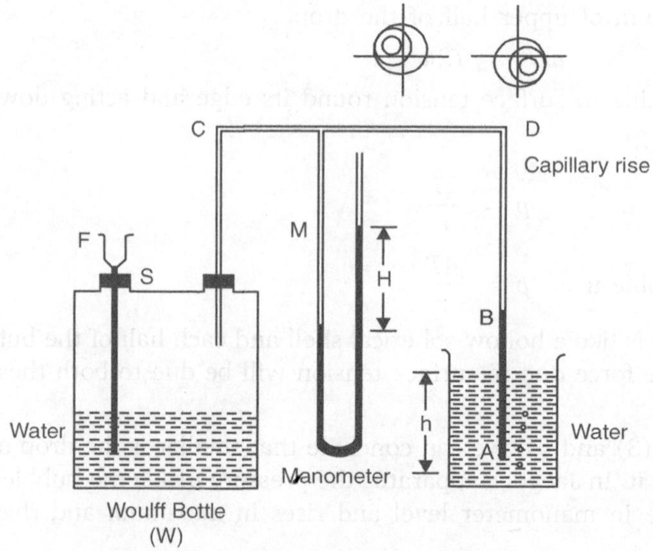

Fig. 9.3. *Experimental arrangement.*

PROCEDURE

1. Make arrangements as shown in Fig. 9.3.
2. Open the stopcock of the dropping funnel slightly so that water slowly falls into the bottle.

3. Adjust the rate of flow of water into the bottle so that a bubble is formed at the end A of the capillary in about 10 seconds. The pressure indicated by the manometer rises and becomes maximum when the bubble has a radius equal to the radius of the orifice at A.
4. Note the maximum difference in the height of the liquid in the two limbs of the manometer. The bubble at this stage is unstable and suddenly breaks away equalizing the level in the limbs of the manometer.
5. Repeat the experiment for the same value of h.
6. Note the depth of the capillary in the experimental liquid with the help of meter scale.
7. Repeat the experiment for different h.
8. Then note the least count of travelling microscope. Take out the capillary and put horizontally as shown in Fig. 9.4 and measure its diameter in two mutually perpendicular directions with travelling microscope (Fig. 9.5).
9. Take mean of the diameter and half it and get radius r of the capillary.
10. Note the temperature of the water.

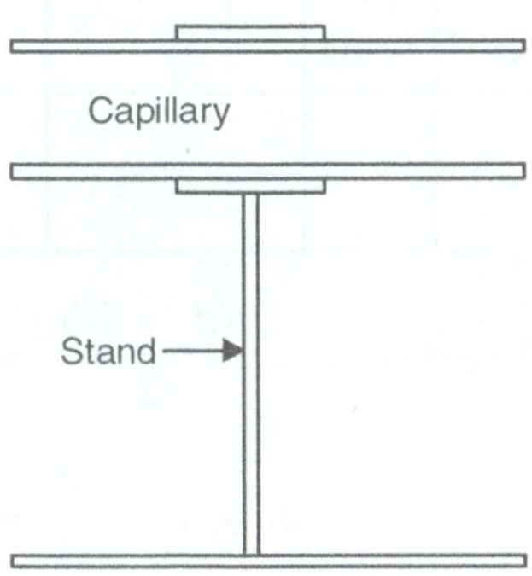

Fig. 9.4

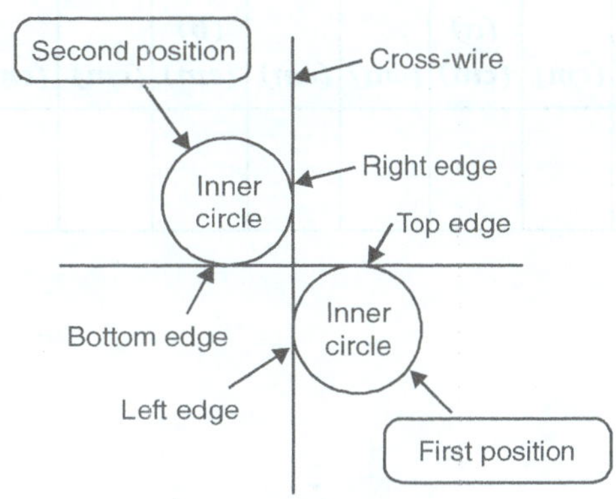

Fig. 9.5

OBSERVATIONS

Table 9.1. Observation for 'H' and 'h'.

S.No.	Depth of capillary in liquid h (in cm)	Manometer reading		Difference $H = x - y$ (cm)	$H - h$ (cm)
		Position of liquid level in one limb x (cm)	Position of liquid level in other limb y (cm)		

Value of 1 division on main scale of travelling microscope, x = ... cm

Total number of divisions on vernier scale, n = ...

$$\text{Least count of travelling microscope} = \frac{\text{Value of 1 main scale division}}{\text{Total number of divisions on vernier scale}}$$

$$= \frac{x}{n} \text{ cm} = \text{....... cm}$$

Table 9.2. Observation for diameter of capillary tube.

Microscope focused on												*Diameter*	
Left edge			*Right edge*			*Top edge*			*Bottom edge*				
M. S. R. (cm)	*V. S. R. (cm)*	*TOTAL (a) (cm)*	*M. S. R. (cm)*	*V. S. R. (cm)*	*TOTAL (b) (cm)*	*M. S. R. (cm)*	*V. S. R. (cm)*	*TOTAL (c) (cm)*	*M. S. R. (cm)*	*V. S. R. (cm)*	*TOTAL (d) (cm)*	$d_1 = a \sim b$ *(cm)*	$d_2 = c \sim d$ *(cm)*

$$\text{Mean diameter, } d = \frac{(d_1 + d_2)}{2} = \text{........ cm}$$

$$\text{Radius, } r = \frac{d}{2} = \text{........ cm}$$

- Room temperature = ... °C
- Density of water = ... gm/cc
- Gravitational acceleration, g = 980 cm/s^2

CALCULATION

Surface tension of water

$$T = (H - h)\frac{rg}{2} = \text{..................... dyne/cm}$$

RESULT

Surface tension of water at a temperature of .... °C = ... dynes/cm

PRECAUTIONS

Please refer to Experiment No. 8.

CONCLUSION

VIVAVOCE

1. What is your object ? Define surface tension ?
2. What is the amount of excess pressure inside the bubble ?
3. Which method of measuring surface tension is more accurate ?
4. What do we measure by manometer ?
5. What will happen if a capillary tube is immersed in water in an inclined way ?
6. Do capillary tube having same radius but made up of the different material show the same rise of water in them ?
7. Will the increase in radius of the capillary tube show any change in observations ?
8. If the tube is smoothened using oil or glycerin how will the observation change ?
9. What is buoyant force ?
10. Why is kerosene oil spilled over small puddles to kill mosquitoes ?
11. What is the maximum height to which water can rise in a capillary tube ?
12. Why does water rise in a capillary tube when immersed in water ?
13. What is the cause of surface tension in the direction of force ?
14. What is cohesive and adhesive force ?
15. Explain effect of impurities and temperature on Surface Tension.
16. Why we measure diameter of capillary tube in two perpendicular directions ?

❑❑❑

VISCOSITY

Viscosity of a liquid is a measure of its inability to flow, and this is measured in Nsm^{-2} (*SI* units) or poise (p).

$$1 \text{ p} = 1 \text{ Nsm}^{-2}$$

Fluids that flow readily, such as water or gasoline, have smaller viscosity than do "thick" liquids such as honey or water oil.

FACTORS AFFECTING VISCOSITY

1. **Size of Molecules.** Viscosity of a liquid having large molecules is high whereas the viscosity of those liquids that have small molecules is low.
2. **Shapes of Molecules.** Spherical molecules provide resistance but oval shaped or disc like molecules provide greater resistance in the flow of liquid. That's why viscosity of liquids having spherical molecules is low.
3. **Inter Molecular Forces.** Liquids having large inter molecular forces have greater viscosity.
4. **Temperature.** Viscosity of liquid decreases with increase in temperature. Because an increase in temperature reduces the forces of attraction between molecules. Such as Lava. Hotter the lava, the more easily it can flow.

Fig. 1 shows a velocity profile for a laminar flow of a viscous fluid in a long cylindrical pipe.

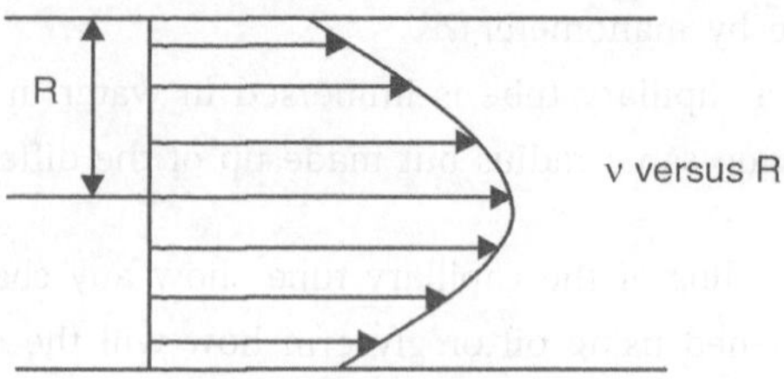

Fig. 1

The motion is like a lot of concentric tubes sliding relative to one another, with the central tube moving fastest and the outer most tube at rest. Viscous forces between the tubes oppose this sliding, so to keep the flow going we must apply a greater pressure at the back of the flow than at the front.

[**Example :** We have to keep squeezing a tube of toothpaste or a packet of ketchup to keep the fluid coming out of its container. Our fingers provide a pressure at the back of the flow that is far greater than the atmospheric pressure at the front of the flow.]

The pressure difference required to sustain a given volume flow rate through a cylindrical pipe of length L and radius R turns out to be proportional to (L/R^4).

CONCLUSION

If we decrease R by one-half, the required pressure increases by $2^4 = 16$; decreasing R by a factor of 0.90 (a 10% reduction) increases the required pressure difference by a factor of (1/0.90) = 1.52 (a 52% increase).

This simple relationship explains the connection between a high cholesterol diet (which tends to narrow the arteries) and high blood pressure. Due to the R^4 dependence, even a small narrowing of the arteries can result in substantially elevated blood pressure and added strain for the heart muscle.

EXPERIMENT—10

AIM

Determination of coefficient of viscosity of a given liquid by Stoke's method.

APPARATUS REQUIRED

Falling sphere viscometer, steel balls of different diameter, stopwatch, screw gauge, thermometer, meter scale, rubber band, and watch glass.

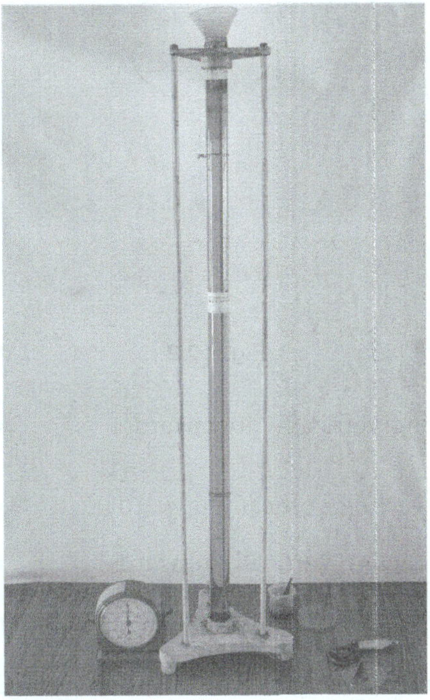

Fig 10.1. *Falling sphere viscometer.*

DESCRIPTION OF THE APPARATUS

The experimental arrangement is shown in Fig. 10.1. The experimental liquid is taken in a one meter long cylindrical glass tube (viscometer) fixed with a thermometer and a funnel at the top end. Diameter of the balls is first measured with screw gauge. The steel balls then rinsed with liquid taken in a watch glass and can be dropped into the viscometer through the funnel. There are three equidistant marks *a, b* and *c* on the tube well below the top (Fig. 10.3).

PRINCIPLE

When a small spherical body moves freely in a viscous medium at rest, a backward dragging force comes into act due to relative motion between different layers of the medium. This backward dragging force is called viscous force. This viscous force, which acts upward, also goes on increasing with the increase in velocity of the moving body. When the effective weight of the spherical body is just equal to the viscous force, the body acquires a constant velocity, which is called terminal velocity.

THEORY

If a small spherical ball of radius r and density σ is allowed to fall freely from rest through a viscous liquid of density ρ and coefficient of viscosity η, then the following three forces act on the ball

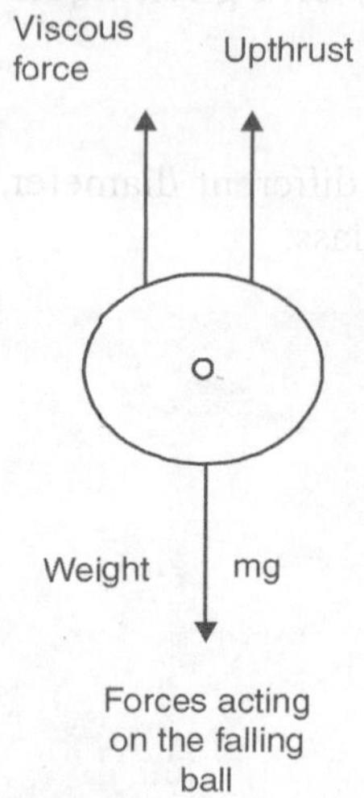

Fig. 10.2

1. Weight of the ball (acting vertically downwards) $= \frac{4}{3}\pi r^3 \sigma g$ ...(1)

2. Upthrust due to liquid (acting vertically upwards) $= \frac{4}{3}\pi r^3 \rho g$ ...(2)

3. Viscous force of liquid (acting vertically upwards) $= 6\pi\eta r v$ ...(3)

From equations (1) and (2), the effective weight of the ball $= \frac{4}{3}\pi r^3 \sigma g - \frac{4}{3}\pi r^3 \rho g$ ...(4)

When ball acquires terminal velocity, then viscous force = Effective weight of the ball

$$6\pi\eta r v = \frac{4}{3}\pi r^3 \sigma g - \frac{4}{3}\pi r^3 \rho g$$

$$6\pi\eta r v = \frac{4}{3}\pi r^3 (\sigma - \rho) g$$

$$\eta = \frac{2}{9}\frac{r^2(\sigma - \rho)g}{v}$$

$$\eta = \frac{2}{9}\frac{r^2(\sigma - \rho)gt}{s}$$

where s is the distance moved by the sphere in time t after attaining terminal velocity and g is gravitational acceleration.

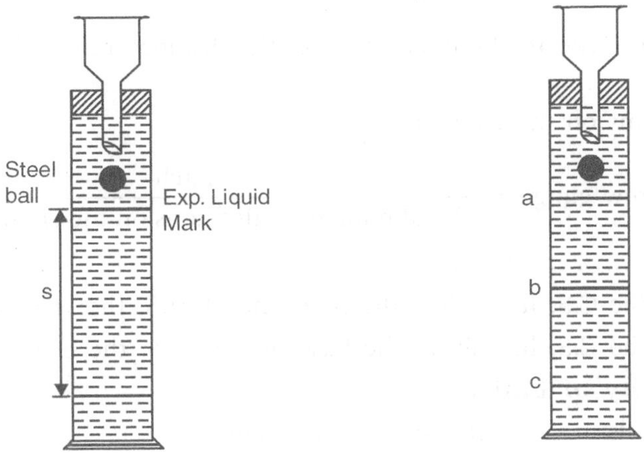

Fig. 10.3. *Experimental arrangement.*

PROCEDURE

1. Firstly calculate the least count of stopwatch.
2. Take a number of steel balls and measure their diameters with a screw gauge.
3. Set the viscometer as shown in Fig.10.3, and tie three rubber bands well below the liquid column as a, b and c equidistant from each other *i.e.,* $ab = bc$.
4. Drop one ball centrally through the liquid along the axis of the tube. When the balls reach the point a, start the stopwatch and note the time taken by it to cover the distance ab and then bc. If the time intervals are equal the ball has acquired the terminal velocity. Find the mean of transit time between ab and bc.
5. Repeat the experiment with other balls of different diameters.
6. Measure the density of the liquid.
7. Note down the temperature of the water.

OBSERVATIONS

Table 10.1. To find the radius of the ball.

Ball No.	*Main scale reading (M.S.R.) (cm)*	*Circular scale reading (C.S.R.) = number of division coinciding the base line × least count (cm)*	*Diameter (M.S.R. + C.S.R.) (in cm)*	*Radius 'r' (in cm)*
A				
B				
C				
D				

$$\text{Pitch} = \frac{\text{Distance travelled on the pitch scale}}{\text{Number of rotations of circular head}} = \ldots\ldots\ldots\ldots \text{ cm}$$

(If 2 rotations are given to the circular scale, the distance moved is 1 mm, the Pitch = ½ = 0.5 mm)

Number of divisions on the circular scale = ...

$$\text{Least count of screw gauge} = \frac{\text{Pitch}}{\text{Total number of divisions on the circular scale}} = \ldots\ldots\ldots \text{ cm}$$

Zero error = ± .... cm

I. If the zero of the C.S. lies below the base line of M.S. the error is positive.

II. If the zero of the C.S. lies above the base line of M.S. the error is negative.

➤ **To find density of balls.**

1. Mass of the balls $m_1 = \ldots\ldots\ldots\ldots$ gm
2. Radius of the ball $r = \ldots\ldots\ldots$ cm
3. Volume of balls $V_1 = \frac{4}{3}\pi r^3 = \ldots\ldots\ldots \text{ cm}^3$
4. Density of the ball $\sigma = \frac{m_1}{V_1} = \frac{m_1}{\frac{4}{3}\pi r^3} = \ldots\ldots\ldots\ldots \text{gm/cm}^3$

➤ **To find density of oil.**

1. Mass of oil $m_2 = \ldots\ldots\ldots\ldots$ gm
2. Volume of the oil $V_2 = \ldots\ldots\ldots$ ml $= \ldots\ldots\ldots$ cc
3. Therefore density of oil $\rho = \frac{m_2}{V_2}$ ....gm/cc

Table 10.2. To measure the transit time and distance.

Balls		***Distance between consecutive threads***		***Transit time***	***Mean time***	***Terminal velocity***
		s (cm)	***Mean s (cm)***	***t (sec)***	***t (sec)***	***v = s/t (cm/sec)***
A	*(i)* *(ii)*					
B	*(i)* *(ii)*					
C	*(i)* *(ii)*					
D	*(i)* *(ii)*					

Mean $v = s/t = \ldots\ldots\ldots\ldots\ldots\ldots$ cm/sec

➤ Room temp. = ... °C

➤ Value of gravitational acceleration, $g = 980 \text{ cm/sec}^2$

CALCULATIONS

The co-efficient of viscosity of the given oil, $\eta = \frac{2}{9}\frac{r^2(\sigma - \rho)gt}{s}$ = ... Poise

RESULT

The co-efficient of viscosity of the given oil at .... °C temp, η = ... Poise

PRECAUTIONS

1. Before the ball is allowed to fall through the liquid, it must be immersed in the experimental liquid.
2. The ball should fall centrally; it should not touch the walls of the tube.
3. Balls should be small in size.

CONCLUSION

VIVAVOCE

1. What do you understand by viscosity of a fluid ?
2. Define co-efficient of viscosity.
3. What are the units and dimensions of co-efficient of viscosity ?
4. From where does the viscous force come here into play ?
5. Does the viscosity vary with temperature ? If yes, how ?
6. What is streamline and turbulent flow ?
7. Define Reynolds's number.
8. Define co-efficient of viscosity.
9. How it depends on radius of the capillary tube ?
10. What is velocity gradient ?
11. How does the viscosity vary with temperature ?
12. On what factors does the viscous force of a liquid in streamline flow depend ?
13. State the SI and CGS units of co-efficient of viscosity. How are they related ?
14. Is it necessary to maintain a constant pressure difference across the capillary tube ?
15. Can we use capillary of large diameter in this experiment ?

❑❑❑

Experiment—11

AIM

To determine the coefficient of viscosity of water by Poiseuille's method.

APPARATUS REQUIRED

A capillary tube of uniform cross-sectional area, a graduated cylinder, a constant level water tank attached on a vertical stand, a stopclock, manometer, travelling microscope, meter scale and a thermometer.

Fig. 11.1. *Poiseuille's apparatus.*

DESCRIPTION OF THE APPARATUS

The experimental arrangement is shown in Fig. 11.1. A water reservoir is attached on a vertical stand which can be easily moved up or down. The reservoir has three openings at its base to which rubber tubes are fitted. From the inlet tube, water is allowed from a tap to run into the tank and its flow is adjusted at such a rate that the level of water in the tank is maintained constant. The water enters through the inlet and goes to the capillary tube. A capillary tube of about 30–40 cm length and of uniform bore (0.5 mm in diameter) is connected horizontally (as shown in Fig. 11.1) to the outlet tube of the tank. The excess water overflows through the central tube of the tank. This ensures a constant level of water in tank. Two rubber tubes are connected at each end of the capillary tube which are attached to the vertical limbs of the manometer and a constant pressure difference across the ends of capillary tube is directly measured by the difference in liquid levels in the limbs of manometer. By altering the height of the tank, the flow of water through the capillary is adjusted, such that the liquid leaves the end of the tube in a trickle, which is collected, in a measuring cylinder.

PRINCIPLE

When a layer of a fluid slips or tends to slip on another layer in contact, the two layers exert tangential forces on each other. The directions are such that the relative motion between the layers is opposed. This is called viscosity of the fluid. Because of this property, in steady flow of a fluid through a narrow cylindrical tube, the layer in contact with the wall of the tube remains at rest and the layers away from the wall move fast. The force of viscosity between two layers of a fluid is proportional to the velocity gradient in the direction perpendicular to the layers. Also the force is proportional to the area of the layer. The flow of a liquid remains streamline so long as its velocity is less than critical velocity $V_c = k\eta/\rho r$. Where η and ρ are the coefficient of viscosity and density of the fluid. r is the radius of the cylindrical tube.

THEORY

Let AB be a capillary tube of length l and radius r. A liquid (water) is flowing through this capillary by applying a constant pressure difference P between the ends of the tube. The flow of the liquid is streamline, steady in the tube and parallel to the axis of the tube (Fig. 11.2). The velocity of the layer is zero which is in contact with the side of the tube and velocity increases in a regular manner towards axis of the tube. Consider a cylindrical layer of radius x. Let the velocity of the liquid be v at a distance x.

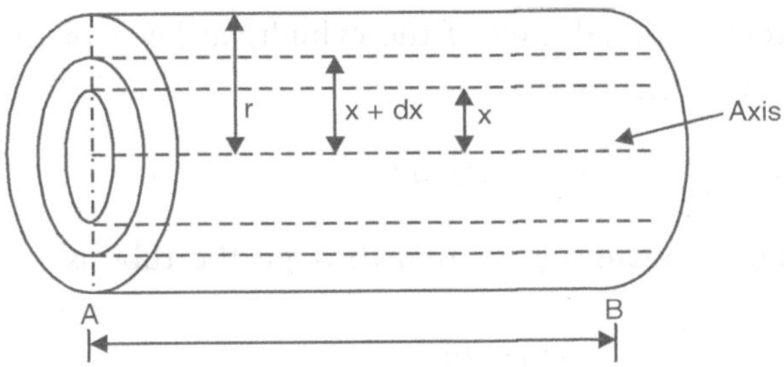

Fig. 11.2

According to Newton's law of viscous flow, the retarding viscous force

$$F = -\eta A \frac{dv}{dx} \qquad \text{...(1)}$$

where $\frac{dv}{dx}$ is velocity gradient in a direction perpendicular to the layer and A is surface area of the layer. Negative sign indicates the direction of this force which acts in a direction opposite to the movement of layer of fluid. Since $A = 2\pi xl$

So, from equation (1)

$$F = -\eta . 2\pi xl \frac{dv}{dx} \qquad \text{...(2)}$$

Now, since P is the pressure difference between two ends of the capillary tube, the force driving the liquid forward is

$$= P.\pi x^2 \qquad \text{...(3)}$$

In steady state, there is no acceleration of the liquid, so the forward force must be equal to the retarding viscous force. From equations (1) and (3)

$$P.\pi x^2 = -\eta.2\pi xl\frac{dv}{dx}$$

or, $$dv = -\frac{P}{2\eta l}.xdx$$

on integrating above equation, we get

$$v = -\frac{P}{2\eta l}.\frac{x^2}{2}+C$$

where, C is a constant of integration. The velocity of the layer in contact with fixed surface is zero *i.e.,* when

$$x = r; \quad v = 0 \quad \text{and} \quad C = \frac{P}{2\eta l}.\frac{r^2}{2}$$

Therefore, $$v = \frac{P}{4\eta l}\left(r^2 - x^2\right)$$

This is the expression for velocity of the liquid at a distance x from the axis of the tube. Now, let us consider that the layer having inner radius x, has thickness dx so that the volume dQ of the liquid flowing per second through it, is

dQ = velocity × cross-sectional area of the cylindrical layer of the thickness dx

$$= v \times 2\pi x.dx$$

$$= \frac{P}{4\eta l}\left(r^2 - x^2\right)\times 2\pi x.dx$$

Thus, the total volume flowing per second through the tube is

$$Q = \int_0^r \frac{P}{4\eta l}\left(r^2 - x^2\right)\times 2\pi x dx$$

Solving above equation, we get

$$Q = \frac{P\pi r^4}{8\eta l} \qquad ...(4)$$

The atmospheric pressure P

$$P = \rho gh$$

where, ρ is the density of liquid, g is gravitational acceleration, and h difference of liquid levels in two limbs of the manometer. Hence, from equation (4)

$$Q = \frac{\pi r^4.\rho gh}{8\eta l}$$

or, $$\eta = \frac{\pi r^4.\rho gh}{8Ql}.$$

FORMULA

Coefficient of viscosity of the liquid, $\eta = \dfrac{\pi r^4.\rho gh}{8Ql}$

where η = Coefficient of viscosity,

l = Length of the capillary tube,

r = Radius of the capillary tube,

Q = Volume of liquid per second through capillary,

ρ = Density of liquid,

g = Acceleration due to gravity,

h = Difference of height of liquid in the manometer.

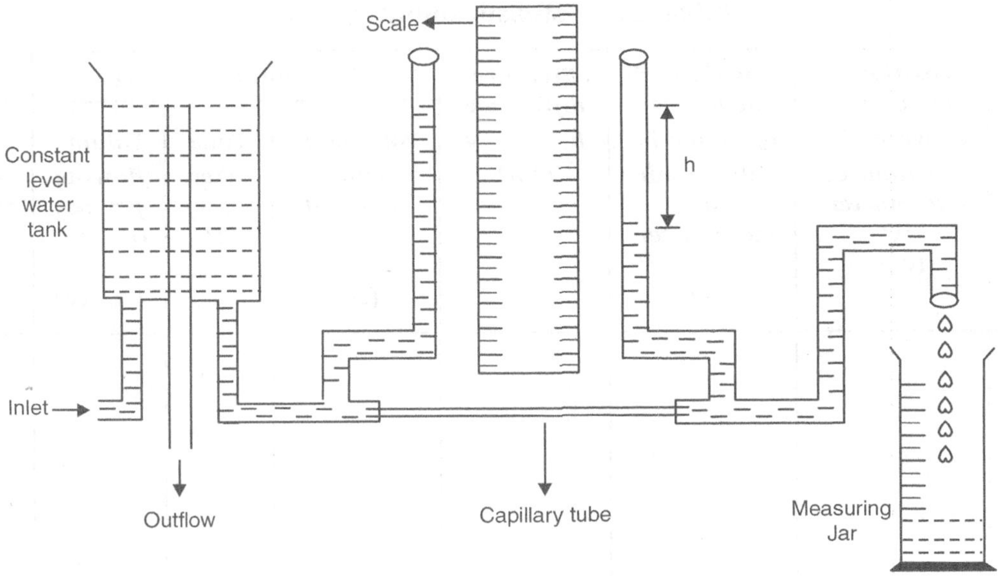

Fig. 11.3. *Experimental arrangement.*

PROCEDURE

1. Arrange the apparatus as shown in Fig. 11.3. Capillary should be exactly horizontal so that the flow of liquid in the capillary is not under gravity.
2. Now enter the water through the opening 1 in the constant level water tank which starts flowing out of capillary through the opening 3 and the excess water above the constant height in reservoir flows out in the sink through the opening 2.
3. Adjust the height of water tank on the vertical stand and the amount of water entering the capillary from the tank in such a manner that water comes out drop by drop from the end of the capillary and the meniscus of water in the two limbs of manometer remain constant. No air bubble should be there in the capillary and the limbs of manometer.
4. When water flows out of the capillary tube steadily and drop by drop, note down the difference of water levels in the manometer which gives h.
5. Collect sufficient quantity of water trickling from the end of the capillary tube in a graduated jar for a definite time interval and find out its volume. Dividing total volume of liquid collected by total time, Q is obtained.

6. Now alter the pressure difference by raising or lowering the water tank. Repeat the experiment three-four times with various values of h and Q. Calculate a mean value of h/Q. Note down the temperature of water.
7. Finally measure the length and radius of the narrow cylindrical tube. Substitute the mean value of h/Q, radius 'r' length 'l' in the formula, we get the value of η.

OBSERVATIONS

Least count of stop watch = ... sec.

Table 11.1. Measurement of h/Q.

S.No.	Position of meniscus of water in one limb of manometer 'x' (cm)	Position of meniscus of water in other limb of manometer 'y' (cm)	Pressure difference $h = x - y$ (cm)	Determination of Q			$\frac{h}{Q}$ (cm. sec/cc)
				Volume of water collected 'V' (cc)	Time taken 't' (sec)	Volume flowing per sec $Q = V/t$ (cc/sec)	

Mean $\frac{h}{Q}$ = ... cm.sec/cc

- Length of capillary tube = ... cm.
- Temperature of water = ... °C
- Diameter of the capillary tube = ... cm
- Density of the water ρ = ... gm/cc
- Gravitational acceleration g = ... cm/sec^2

CALCULATIONS

Coefficient of viscosity of water, $\eta = \frac{\pi r^4 . \rho g h}{8Ql}$ = ... Poise

RESULT

Coefficient of viscosity of water, η (Room temperature = ... °C) = ... Poise

PRECAUTIONS

1. The bore of the capillary tube should be about 0.5 mm.
2. The capillary tube should be uniform in section.
3. The capillary tube should be placed horizontally.
4. The diameter of the capillary tube should be measured accurately at both the ends.
5. While taking observations from microscope, backlash error should be avoided.
6. The pressure between the ends of the capillary tube should be small so that liquid may leave the tube in a trickle.
7. To measure Q, a graduated cylinder of small least count should be used and sufficient quantity of water should be collected in it.
8. The pressure difference between the ends of the tube should be constant. Hence the level of water in the constant level water tank should be maintained constant.
9. Temperature of water should be measured accurately with a sensitive thermometer.

CONCLUSION

VIVAVOCE

Please refer to Experiment No. 10.

❑❑❑

EXPERIMENT—12

AIM

To determine the mechanical equivalent of heat '*J*' by Callender and Barne's method.

APPARATUS REQUIRED

Callender and Barne's continuous flow calorimeter, constant level water flow tank, 12 V battery eliminator, ammeter and voltmeter, two thermometers, stopwatch, graduated cylinder, rheostat, key and connecting wires.

Fig. 12.1. *Callender and Barne's apparatus.*

DESCRIPTION OF THE APPARATUS

It consists of a heating coil made of nichrome or manganin wire (high or moderate resistant wire), enclosed inside a horizontal glass tube along its length. The glass tube is surrounded by a wooden box and insulator fiber to minimize heat losses. Two thermometers are inserted in the glass tube to measure the incoming and outgoing temperature of the water. Continuous flow of water at a steady rate can be maintained through the glass tube with the help of constant level water tank. The water tank is a metal vessel having three openings at the bottom. One of the openings is connected to the inlet tap, the middle one to the sink, and the other to the outlet end. The flow of water can be maintained by means of stopcock attached to the output end. The heated water through the glass tube can be collected in a measuring cylinder. A power supply (or a battery), rheostat, and ammeter are connected in series between the ends of the helical coil, and a voltmeter is connected in parallel across the coil.

PRINCIPLE

When a steady current flows through the tube, which encloses the heating coil and steady flow of water is maintained through the tube, the amount of electrical energy supplied during a particular time, is consumed in heating the amount of water which flows through the tube during the same time interval and a small amount of heat is lost by radiation etc. to the surrounding during that interval.

THEORY

The mechanical equivalent of heat J is the amount of mechanical work in joules required to raise the temperature of 1 g of water by 1°C and

$$J = \frac{W}{H}$$

or $$H = \frac{W}{J} \quad ...(1)$$

where H is the heat produced in calorie and W is the work in joule.

If current I flows through the coil at potential V for time t sec, then

$$W = VIt \text{ joule}$$

∴ From equation (1), heat produced in the coil

$$H = \frac{VIt}{J} \text{cal} \quad ...(2)$$

But according to the principle of calorimetry, heat taken by water in t sec

$$H = ms(\theta_2 - \theta_1) \text{ cal} \quad ...(3)$$

where, s is the mean specific heat of water, m is the mass of water collected in steady state and $(\theta_2 - \theta_1)$ is the temperature difference of incoming and outgoing water.

In steady state, heat produced in the coil = heat taken by the water + heat loss

∴ From equations (2) and (3),

$$\frac{VIt}{J} = ms(\theta_2 - \theta_1) + h \quad ...(4)$$

where h is the heat lost by radiation.

In the first set of observation, current I_1 is passed through the heating coil at a potential V_1. Steady state is achieved, at the temperature difference $(\theta_2 - \theta_1)$ of incoming and outgoing water. If m_1 gm of water flows in time t sec in the steady state, then from equation (4),

$$\frac{V_1 I_1 t}{J} = m_1 s(\theta_2 - \theta_1) + h_1 \quad ...(5)$$

In the second set of observation, I_1, V_1 and m_1 are changed to I_2, V_2 and m_2 while keeping the temperature difference constant. Then we can write,

$$\frac{V_2 I_2 t}{J} = m_2 s(\theta_2 - \theta_1) + h_2 \quad ...(6)$$

Subtracting equation (5) from equation (6),

$$\frac{(V_2 I_2 - V_1 I_1)t}{J} = (m_2 - m_1)(\theta_2 - \theta_1)s + (h_2 - h_1) \quad ...(7)$$

For all practical purpose, we consider $h_1 = h_2$,

$\therefore$ Equation (7) becomes

$$\frac{(V_2I_2 - V_1I_1)t}{J} = (m_2 - m_1)(\theta_2 - \theta_1)s$$

or
$$J = \frac{(V_2I_2 - V_1I_1)t}{(m_2 - m_1)(\theta_2 - \theta_1)s} \quad ...(8)$$

Taking specific heat of water s = 1 cal/gm °C,

$$J = \frac{(V_2I_2 - V_1I_1)t}{(m_2 - m_1)(\theta_2 - \theta_1)} \text{ joule/cal} \quad ...(9)$$

Thus, the mechanical equivalent of heat J can be calculated from the above expression.

> **Note.** *Special care must be taken that, the temperature difference between the out going and incoming water should be about 4°C to 5°C and this value should be same in both the sets. This is necessary to avoid error due to radiation losses. However, for some unavoidable reasons, if it is not possible to keep same temperature difference for both sets, then value of J can be calculated separately for each set using following formula and mean should be taken.*

$$J_1 = \frac{V_1I_1t}{m_1(\theta_2 - \theta_1)} \text{ joule/cal}$$

and
$$J_2 = \frac{V_2I_2t}{m_2(\theta_2 - \theta_1)} \text{ joule/cal}$$

So
$$J = \frac{J_1 + J_2}{2}.$$

FORMULA USED

The mechanical equivalent of heat

$$J = \frac{(V_2I_2 - V_1I_1)t}{(m_2 - m_1)(\theta_2 - \theta_1)}$$

where, m_1 = mass of water flowing in time t sec through the tube when current I_1 ampere passes through the coil at a potential difference V_1 volt, m_2 = mass of water flowing in same time t sec through the tube when current I_2 ampere passes through the coil at a potential difference V_2 volt, $(\theta_2 - \theta_1)$ = difference in temperature of outgoing and incoming water in the steady state.

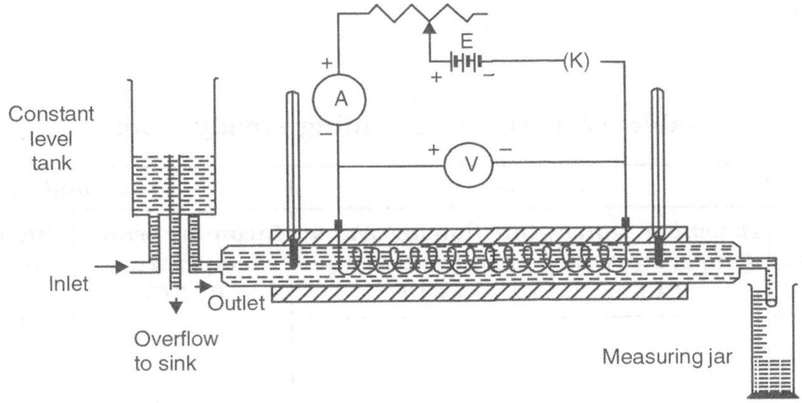

Fig. 12.2. *Experimental arrangement.*

PROCEDURE

1. Firstly make connections as shown in Fig. 12.2.
2. Then allow water to flow through the glass tube slowly and steadily by means of constant level water tank.
3. Now, for first set of observations, the rheostat is adjusted such that a potential difference of 6 volts is applied, across the two ends of the heater coil (kept inside the glass tube). Care should be taken that this potential difference remains constant during the observations.
4. Then, note the readings of thermometers T_1 and T_2 at a regular interval (say 5 minutes). Initially the temperature of incoming water θ_1 remains constant but the temperature of outgoing water θ_2 begins to increase and after some times, it becomes constant. This state is called the steady state. Now wait for some time and check that the steady state has been truly achieved (at least three consecutive readings).
5. Adjust the flow of water by means of a stopcock provided with constant level water tank so that the temperature difference $(\theta_2 - \theta_1)$ of out flowing and incoming water becomes nearly 4 to 5°C.
6. After achieving steady state, start the stopwatch and collect the out flowing water in a graduated cylinder for a definite time t (about 30 sec).
7. For second set of observations, apply the potential difference across the two ends of heater coil V_2 = 10 volt by rheostat. This should again be kept constant during the observations.
8. Now adjust the flow of water by the means of stopcock, so that in the steady state the temperature difference $(\theta_2 - \theta_1)$ of outflowing and incoming water remains the same as in the first set.
9. In this condition, note the reading of voltmeter V_2 and reading of ammeter I_2.
10. Then start the stopwatch and collect the outflowing water in a graduated cylinder for same time t (about 30 sec).
11. Now, measure the volume of collected water and then find its mass m_2 by dividing density of the water.

OBSERVATIONS

Least count of stopwatch = ... sec

Table 12.1. For ascertaining steady state.

		First set		*Second set*	
S.No.	*Time (minutes)*	*Incoming water temperature θ_1 (°C)*	*Out flowing water temperature θ_2 (°C)*	*Incoming water temperature θ_1 (°C)*	*Out flowing water temperature θ_2 (°C)*
1.	0				
2.	5				
3.	10				
4.	15				
5.	20				
6.	25				
7.	30				
8.	35				
...	...	Value of θ_1	Value of θ_2	Value of θ_1	Value of θ_2
...	...	Same value as above	Same value as above	Same value as above	Same value as above
...	...	..., ...	..., ...	..., ...	..., ...
...	...	Steady state achieved	Steady state achieved	Steady state achieved	Steady state achieved

- **When the steady state has been attained :**

		First set	**Second set**
1.	Reading of voltmeter	V_1 = ... volts	V_2 = ... volts
2.	Reading of ammeter	I_1 = ... ampere	I_2 = ... ampere
3.	Temp. of incoming water	θ_1 = ... °C	θ_1 = ... °C
4.	Temp. of out-flowing water	θ_2 = ... °C	θ_2 = ... °C
5.	Time of collection of water	t = ... sec	t = ... sec
6.	Volume of water collected	= .... ml	= .... ml
7.	Mass of water collected	m_1 = ... gm	m_2 = ... gm

(Taking density of water = 1 gm/cc)

CALCULATIONS

The mechanical equivalent of heat

$$J = \frac{(V_2 I_2 - V_1 I_1)t}{(m_2 - m_1)(\theta_2 - \theta_1)} = ...\text{joule/cal}$$

RESULT

The value of mechanical equivalent of heat J = ... joule/cal

PRECAUTIONS

1. The heater coil within the glass tube should be in the form of helical so that the water flowing through the tube gets automatically stirred.
2. Flow of water should be slow and steady.
3. All the readings should be taken strictly after steady state obtained.
4. There must be no air bubble in the tube.
5. All the fittings should be airtight and there should be no leakage of water.
6. The rate of flow of water should be so adjusted that the difference in the readings of the two thermometer (θ_2 – θ_1) in the steady state should be nearly 4 to 5°C.

CONCLUSION

VIVAVOCE

1. Define mechanical equivalent of heat.
2. What are the units of J ?
3. What is the standard value of J ?
4. What is the Joule's law on heating effect of current ?
5. How much work is done in heating by means of current ?
6. Why the heater coil is taken in the helical form ?
7. Why observations are to be taken only after steady state is achieved ?
8. Why the flow of water should be constant and continuous ?
9. Why do you keep the constant temperature difference in both sets ?
10. What will happen to the value of J if heater coil of half the length is taken ?
11. Of what material the heater coil is made ? Can you use copper wire as coil in this experiment ?
12. Is mechanical equivalent of heat J is a physical constant like gravitational acceleration g ?

❑❑❑

EXPERIMENT—13

AIM

To determine the thermal conductivity of a bad conductor by Lee's disk method.

APPARATUS REQUIRED

Lee's disk apparatus, boiler, stopwatch, screw gauge, vernier callipers, and two sensitive thermometers.

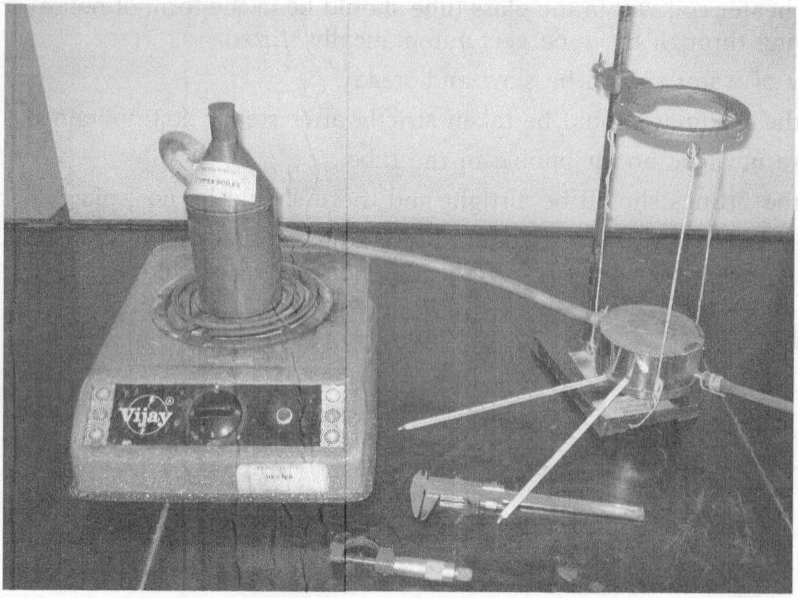

Fig. 13.1. *Lee's disk setup.*

DESCRIPTION OF THE APPARATUS

The main parts of a Lee's disk apparatus are the two metallic disks *A* and *B* having the same diameter (as levelled in Fig. 13.2). The upper disk *A* is a hollow disk, which serves as the steam chamber. The steam chamber is having the openings for the incoming and outgoing of the steam. The lower disk *B* is the experimental disk. Both *A* and *B* disks are having arrangement for placing thermometer in them. The specimen *D* is in the shape of a thin disk having the diameter of the disks *A* and *B* and is well pressed between the two disks *A* and *B*. A little amount of glycerin is applied to both the surfaces of the specimen to make proper thermal contact with both *A* and *B*. The boiler is used for producing the steam to heat the upper steam chamber. Two sensitive thermometers are placed properly in the place provided for them to measure the temperature of the disks.

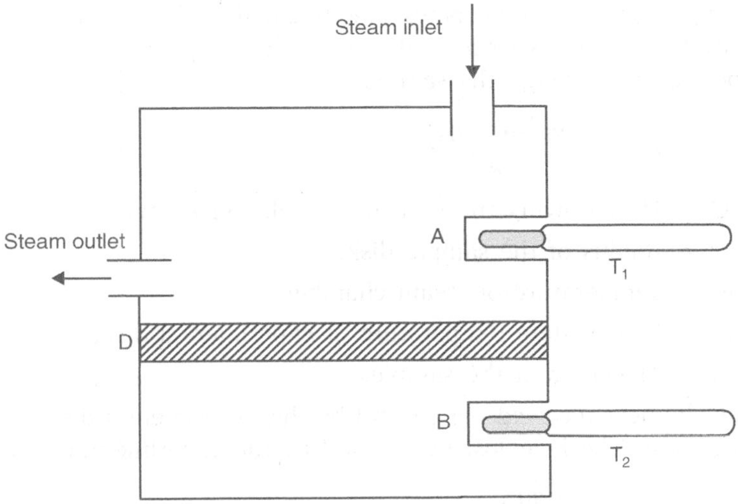

Fig. 13.2. *Experimental arrangement.*

PRINCIPLE

When two sides of a bad conductor specimen is kept at two different temperatures, then the specific amount of heat flows through it and the amount of heat flowing per second :

(*i*) Is proportional to the area of cross-section of the specimen.

$$Q \propto A \qquad \text{(Here } A = \pi r^2\text{)}$$

(*ii*) Is inversely proportional to the thickness of the specimen.

$$Q \propto \frac{1}{d} \qquad \text{(Thickness} = d\text{)}$$

(*iii*) Is proportional to the temperature difference between the two sides.

$$Q \propto (\theta_1 - \theta_2)$$

Therefore,

$$Q \propto \frac{\pi r^2(\theta_1 - \theta_2)}{d}$$

or

$$Q = K\frac{\pi r^2(\theta_1 - \theta_2)}{d}.$$

So by knowing the amount of heat flowing per second, radius and the thickness of the bad conducting sample disk, temperature difference across it, the coefficient of thermal conductivity can be determined.

THEORY

The steam chamber is held at boiling point of the water and the experimental disk is at room temperature initially. As the heat starts flowing through the sample the temperature of the experimental disk starts increasing, consequently the amount of flow of heat decreases due to decrease in the temperature difference across sample. At the same time the radiation loss from the experimental (lower) disk increases with increase of its temperature. At a certain stage the heat received per second by the experimental disk becomes exactly equal to the heat lost per second by it and the temperature variation of the experimental (lower) disk stops, as it is the steady state. Hence the heat flowing per second through the sample

$$Q = K\frac{\pi r^2(\theta_1 - \theta_2)}{d} \qquad ...(1)$$

where,
Q = Heat flowing through the sample per second,
r = Radius of the sample disk,
θ_1 = Temperature of steam chamber,
θ_2 = Steady state temperature of experimental disk,
d = Thickness of the sample.

At steady state the heat received per second by the experimental disk must be equal to the heat lost per second by it. The heat lost per second by the experimental disk at θ_2 is given by

$$Q = ms\left(\frac{d\theta}{dt}\right)_{\theta_2} \qquad ...(2)$$

where,
m = Mass of the experimental disk,
s = Specific heat of the disk,
$\left(\frac{d\theta}{dt}\right)_{\theta_2}$ = Rate of change of temp at steady state temp.

But total surface area of the experimental disk cannot radiate heat as the upper surface is covered by the sample so to get the exact amount of heat getting radiated per second the above equation must be multiplied by the fraction of the surface, which is

Surface area radiating heat/ total surface area of the disk

$$= \frac{(\pi r^2 + 2\pi rl)}{(2\pi r^2 + 2\pi rl)} = \frac{(r + 2l)}{(2r + 2l)}$$

where,
l = The thickness of the experimental disk.

So the actual amount of heat radiated per second by the experimental disk is

$$Q = ms\left(\frac{d\theta}{dt}\right)_{\theta_2}\left(\frac{(r+2l)}{2r+2l}\right) \qquad ...(3)$$

From equations (1) and (2)

$$K\frac{\pi r^2(\theta_1 - \theta_2)}{d} = ms\left(\frac{d\theta}{dt}\right)_{\theta_2}\left(\frac{(r+2l)}{2r+2l}\right)$$

or

$$K = ms\left(\frac{d\theta}{dt}\right)_{\theta_2}\left(\frac{(r+2l)}{2r+2l}\right)\frac{d}{\pi r^2(\theta_1 - \theta_2)}$$

$\left(\frac{d\theta}{dt}\right)_{\theta_2}$ can be obtained by getting the slop of the tangent on the cooling curve at the steady state temperature.

PROCEDURE

1. The diameter (2*r*) of the experimental disk (lower disk) is measured with the help of the vernier callipers. The thickness of the sample (*d*) and the thickness (*l*) of the experimental disk (lower disk) is measured carefully with the help of screw gauge. Average of several readings is taken for accuracy in measurement.
2. Mass of the experimental disk is measured with the help of good balance.
3. Keep the three disks as shown in the figure. The experimental disk is the lowermost disk. The sample is kept above it. The radius of the sample must be equal to the radius of the lower disk. The steam chamber is kept at the top of the arrangement.
4. Steam is passed through steam chamber *A* for a pretty long time and the readings of θ_1 and θ_2 are recorded in the Table 13.5 at an interval of 4–5 minutes, until the steady state is reached *i.e.*, the readings of θ_1 and θ_2 become constant and the variation in the readings stop.
5. The sample is now removed and the steam chamber is kept in direct contact with the experimental disk to raise its temperature up to $(\theta_2 + 10)$°C.
6. The steam chamber is now removed and the experimental disk is allowed to cool up to the temperature of $(\theta_2 - 10)$°C. The temperatures of the disk are recorded at a regular interval of time to draw the cooling curve.
7. The cooling curve is plotted with the help of the above observation keeping time in the *x*-axis and temperature in the *y*-axis. The tangent is drawn at θ_2 and its slop is determined to get the value of $\left(\frac{d\theta}{dt}\right)_{\theta_2}$.

OBSERVATIONS

Table 13.1

S.No.	Parameter	Value with unit
1.	Mass of experimental disk	
2.	Specific heat of experimental disk	

Least count of vernier callipers.

Value of smallest div. in M.S. = ...

Number of div. in V.S. = ...

$$\text{Vernier constant of the vernier scale} = \frac{\text{Value of smallest division of main scale}}{\text{Total number of vernier scale divisions}}.$$

Table 13.2. Diameter of experimental disk.

S.No	Main scale reading (cm)	Vernier scale reading (cm)	Total reading (cm)	Mean value of 2r
1.				
2.				
3.				

$$\text{Pitch} = \frac{\text{Distance travelled on the pitch scale}}{\text{Number of rotations of circular head}} = \text{.............. cm}$$

(If 2 rotations are given to the circular scale, the distance moved is 1 mm, the Pitch = $\frac{1}{2}$ = 0.5 mm)

No. of division in the circular scale =

$$\text{Least count of screw gauge} = \frac{\text{Pitch}}{\text{Number of divisions on the circular scale}} = \text{...... cm}$$

Zero error = ± .... cm

I. If the zero of the C.S. lies below the base line of M.S. the error is positive.

II. If the zero of the C.S. lies above the base line of M.S. the error is negative.

Table 13.3. Measurement of thickness of specimen (*i.e.,* cardboard disk *d*).

S.No.	*Main scale reading (cm)*	*Circular scale reading (C.S.D. × L.C.) (cm)*	*Total reading = M.S.R. + C.S.R. (cm)*	*Mean value of d (cm)*
1.				
2.				
3.				

Table 13.4. Measurement of thickness of experimental disk (*l*).

S.No.	*Main scale reading (cm)*	*Circular scale reading (C.S.D. × L.C.) (cm)*	*Total reading = M.S.R. + C.S.R. (cm)*	*Mean value of d (cm)*
1.				
2.				
3.				

Table 13.5. Observation for steady state.

S.No.	*Time (minutes)*	*Temperature θ_1 (°C)*	*Temperature θ_2 (°C)*
1.	0		
2.	5		
3.	10		
4.	15		
5.	20		
6.	25		
...	...		
...	...		
...	...		
...	...	Value of θ_1	Value of θ_2
...	...	Same value as above	Same value as above
...	...	.... ,, ...	.... ,, ...
...	...	Steady state achieved	Steady state achieved

Table 13.6. Observation of cooling of experimental disk.

S.No.	*Time in seconds*	*Temperature in (°C)*
1.	0	$\theta_2 + 10$
2.	30	...
3.	60	...
4.	90	...
5.	120	...
...	...	...
...	...	...
...	...	...
...	...	Up to $\theta_2 - 10$

CALCULATIONS

The value of thermal conductivity of the sample is determined by putting the values of the different parameters in the formula

$$K = ms\left(\frac{d\theta}{dt}\right)_{\theta_2}\left(\frac{(r+2l)}{2r+2l}\right)\frac{d}{\pi r^2(\theta_1-\theta_2)}$$

RESULT

The value of the thermal conductivity of the given bad conductor is .... cal./sec/cm^2.

PRECAUTIONS

1. The steady state temperatures must be recorded when the temperature reading of both the thermometers become constant.
2. Diameter of the sample must be equal to the diameter of the experimental disk.

CONCLUSION

VIVAVOCE

1. Define thermal conductivity.
2. Distinguish between thermal conductivity and thermal diffusivity.
3. What is the meaning of the steady state ?
4. Why is steady state necessary in this experiment ?
5. Define co-efficient of thermal conductivity.
6. What is its unit ?
7. What is the advantage of this method ?
8. What other methods are there to measure the thermal conductivity of bad conductor ?
9. How can you apply this method for a liquid ?
10. Co-efficient of thermal conductivity depends on which parameters ?
11. What happens if we use a thick specimen ?
12. Why and when the sample is kept in direct contact with the experimental disk to raise its temperature up to $(\theta_2 + 10)°C$?
13. What happens to the heat given after the steady state is reached ?

❑❑❑

SOURCES OF LIGHT

There are two main types of sources that can emit light viz **natural sources** of light, and **artificial sources.** Natural light sources are objects that are not made by man. An example of this is the sun, or lightning. However, an artificial light source is an object that emits light that is made by man.

Light source is essential in any optics lab for understanding the behavior of optical phenomena. Here are the details of four **artificial** sources frequently used in lab.

1. Incandescent Bulb (Filament Type)

Mains: 230 V, 50 Hz, and Output Power : 40 W, 60 W, and 100 W etc.

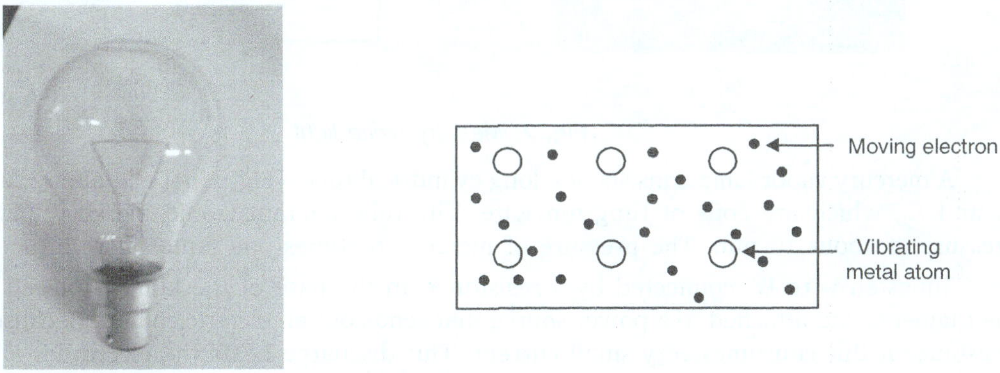

(a) (b)

Fig. 1. *(a) Incandescent bulb, (b) Section of filament of incandescent light bulb.*

An incandescent source means a light source that gives off light when a material is heated enough that it glows white. The electric heating is used here to produce light. A very common example of an incandescent light source is the light bulbs and candles we have in our homes.

In an incandescent light bulb, the electricity goes from the wires into the bulb, which is made up of glass. The electricity then gets forced to go through a very tiny and thin wire called the filament. Here the filament must retain as much of the heat generated as is possible, so that it gets very hot and emits light. It must not melt at such high temp. A strong metal with high melting point such as tungsten (melting point 338°C) is used for making bulb filaments. The filament should be thermally isolated as much as possible, using insulating support, etc. The bulb is usually filled with chemically inactive nitrogen and oxygen gases to prolong the life of filament.

When the supply is given through electrical contacts placed at the bottom of the glass bulb, ionization of metal atoms will takes place. The electrons have a tiny amount of space to go through, so the filament heats up. Moving electrons collide with metal atoms. Vibration of metal atoms increases the temperature and the hot filament given out light. Most of the power consumed by the filament appears as heat, but a small part of it is in the form of light radiated.

Incandescent light bulbs are not very efficient, however, because a large portion of the electricity is converted to heat instead of useful light.

2. Polychromatic Source (Mercury White Light)

Mains : 230 V, 50 Hz, and Output Power : 160 W, λ = 400–800 nm.

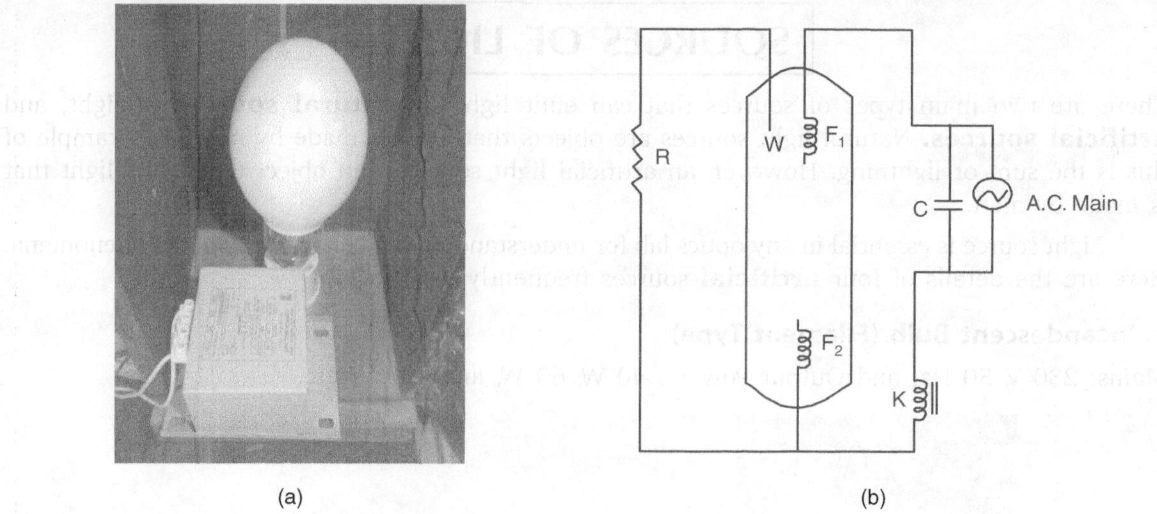

(a) (b)

Fig. 2. *Mercury white light.*

A mercury vapor lamp consists of a long cylindrical tube, [Fig. 2(*b*)] containing two electrodes F_1 and F_2, which are coils of tungsten wire. The tube contains some mercury and argon at a pressure of about 10 mm. The pressure of mercury becomes one atmosphere after sometime.

Tungsten wire W, connected by a resistor R, in the base of the lamp is placed near one of the filaments and attached to a power source that sends out an electric charge. Because of the high resistance R this consumes very small current. This discharge heats the electrodes F_1 and F_2 and the ionization of mercury starts because of which F_1 and F_2 becomes more conducting. Hence the discharge spread between F_1 and F_2.

Once the main discharge between F_1 and F_2 begin to work all the current passes through the tube. Thus, sufficient heat is produced and this vaporizes the mercury. At this stage the discharge is taken over by mercury. Thus the charge causes the mercury vapor inside the tube to give off invisible ultraviolet rays. When the phosphor coating on the outside absorbs the invisible ultraviolet light, it glows and gives off visible light that humans can see. The lamp gives out a pale bluish light and then after a few minutes it attains maximum intensity.

The choke in series with the lamp limits the current (the working voltage is about 180 V). Choke is a coil of very low value of resistance but having large inductance. Choke is preferred over ordinary resistance because in them the loss of energy as heat does not take place, whereas in ordinary resistances there is tremendous loss of energy due to heating effect. Choke consists of an insulated thick copper coil wound over an iron core. In the choke coil a phase difference of about 90° is established between the current and the e.m.f. (large impedance) because of which the power consumed by the lamp becomes zero.

The working temp at this lamp is 600°C. If the working mercury lamp is switched off then it cannot be started again at once because for the electrodes to start their function again, mercury vapors should be first condensed. If mercury lamps are operated at low pressures they emit the radiations corresponding to the ultraviolet region. Fluorescent tubes are more efficient than incandescent light bulbs because they produce less heat than incandescent light bulbs do. However, fluorescent tubes have contents that are poisonous and the tubes are also harder to dispose.

3. Monochromatic Source (Sodium Vapor Lamp)

Mains : 230 V, 50 Hz, and Output Power : 35 W, λ = 5893 Å

This light source requires a special power supply. Connecting directly to the mains may damage the lamp.

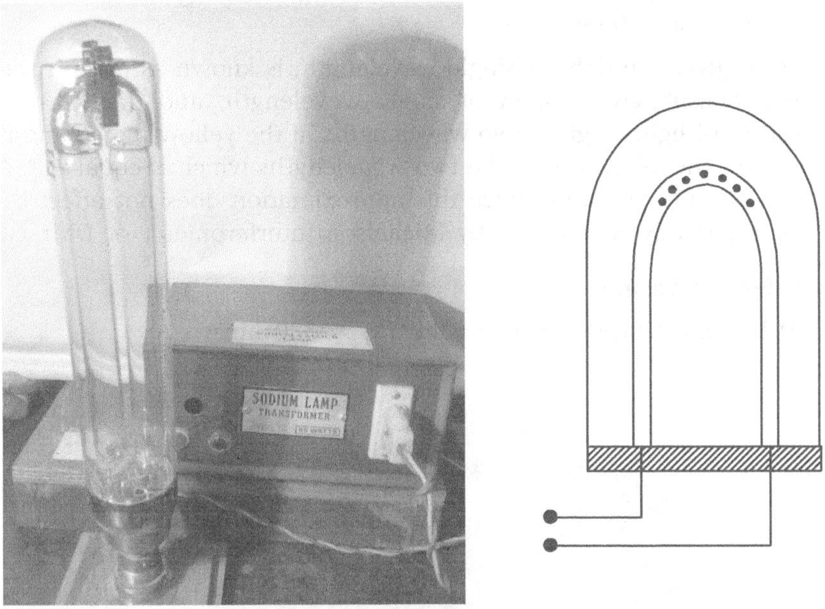

Fig. 3. *Sodium vapor lamp.*

This lamp consists of a U-tube (Fig. 3) with electrode made of tungsten spiral coated with barium oxide. The tube contains rare gases *e.g.*, neon gas under low pressure of about 10 mm and a small quantity of metallic sodium is deposited on the wall of the tube. This discharge tube is enclosed in a unsilvered vacuum jacket to avoid losses of heat. Sodium vapor attacks common glass; therefore all the glass parts of the tube are coated inside with a film of special borosilicate glass.

The working temperature of the lamp is about 250°C. If this temperature is not maintained constant then the vapor pressure and hence the intensity of emitted light will be considerably varied. Sodium vapor lamp works with a voltage of 440 volts. The usual voltage in the mains varies is 230 volts. Therefore, a voltage of 440 volts is obtained by means of a step-up transformer. The lamp is connected to the secondary of a transformer. But the same voltage is not required for maintaining the discharge. A comparatively low voltage of about 130 volts is required for it. Because the ionization potential of sodium is much less than that of neon, therefore, after the moment when discharge starts passing through sodium, a much lower voltage is required to maintain the discharge.

When the lamp is switched on, a voltage of about 440 volts (the lamp is connected to the secondary of a step-up transformer) acts across the tube and discharge passes first through neon gas (because the ignition voltage of neon is quite low) and provides a path for discharge. Therefore, red light is first emitted. Now the tube starts taking current and after a few minutes the voltage falls to about 170 volts, the tube gets heated and metallic sodium in the tube gets vaporized. The vapor pressure of sodium is very low and it is such that no discharge can pass through it unless the temperature is raised (at this temp. sodium becomes liquid and may get localized therefore it is advised not to give any jerk because it may reduce the life of the lamp). Now the lamp starts emitting the light of yellow color.

The lamp emits sodium light because the voltage required to excite the sodium atoms is much lower than that required exciting the neon atoms. Further since the neon atoms lose their energy on collision with sodium atom, hence they do not have much energy to excite themselves. In thus the light emitted is that of sodium.

A source, which gives the light of single wavelength, is known as monochromatic source of light. Sodium lamp doesn't give a light of single wavelength, and then also we call it as a monochromatic source of light. It gives two wavelengths in the yellow region and is called D_1 and D_2 lines of sodium. But the difference in the two wavelengths which is equal to 6 A.U. is so small that they appear to be only one and hence this approximation does not affect the experiments. These lines can be separately seen either by Michelson Interferometer or Diffraction Grating.

4. Laser Source (He-Ne Laser)

Mains : 230 V, 50 Hz, and Output Power : 3 mW, λ = 630 nm

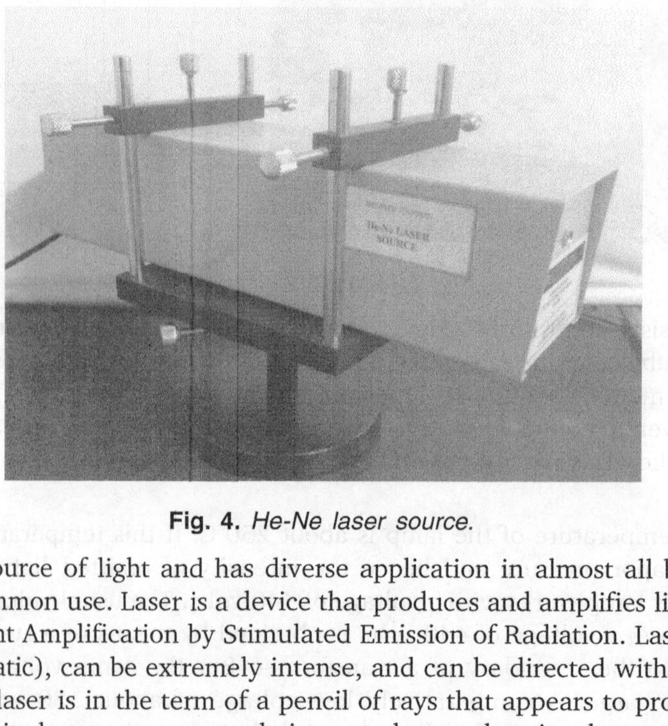

Fig. 4. *He-Ne laser source.*

It is a unique source of light and has diverse application in almost all branches of science, technology and in common use. Laser is a device that produces and amplifies light. The word Laser is an acronym for Light Amplification by Stimulated Emission of Radiation. Laser light is very pure in color (monochromatic), can be extremely intense, and can be directed with great accuracy. The beam coming from a laser is in the term of a pencil of rays that appears to propagate without any divergence. When excited atoms return to their ground state then it releases it energy as light. It can be done in two processes :

(1) It can quickly jump back to the low energy level, giving off the extra energy as light or

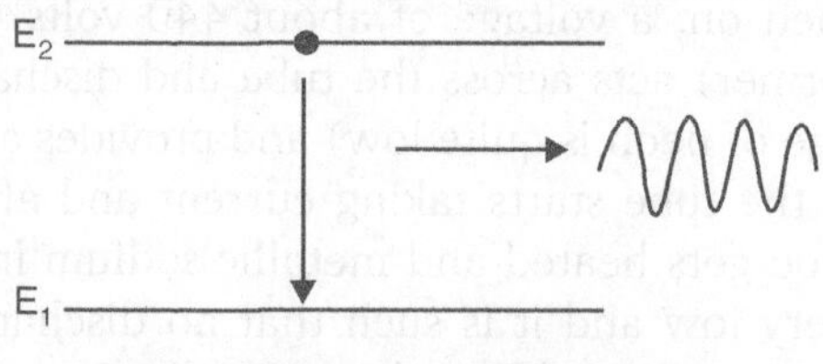

Fig. 5. *Spontaneous emission.*

(2) It may be stimulated by a photon of proper energy emitted by another atom and emit a similar new photon in the process.

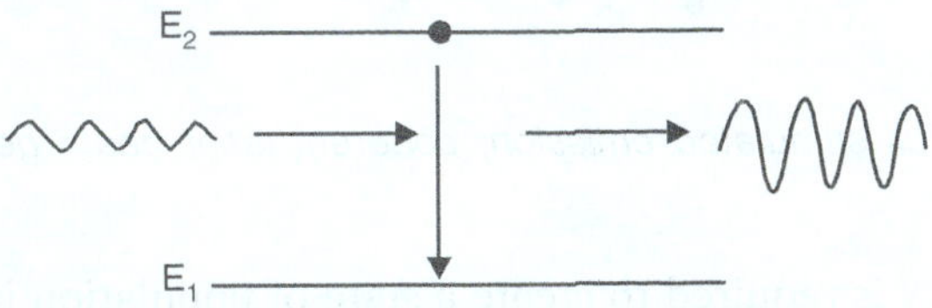

Fig. 6. *Stimulated emission.*

First process is called as spontaneous emission second process is known as stimulated emission. Unlike ordinary source of light, such as a bulb or the sun, where the emission is primarily due to ***spontaneous emission.*** Laser emission uses the principle of ***stimulated emission*** of light.

The three ***main component*** of any laser are the active medium, the pump and the optical resonator.

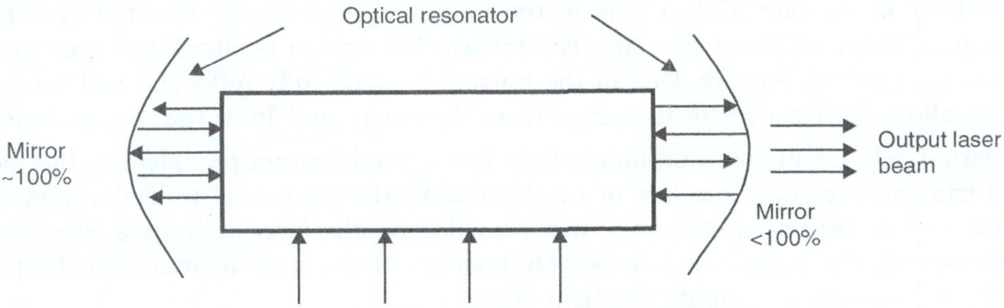

Fig. 7. *Main component of laser.*

The active medium :

It is a material medium of atoms, molecules or ions (in solid, liquid or gaseous form), which is capable of amplifying light waves. Under normal equilibrium condition, there are always a large number of atoms in the lower energy (or ground) state than in the excited energy state. Therefore, a light wave passing through such a collection of atoms would get attenuated.

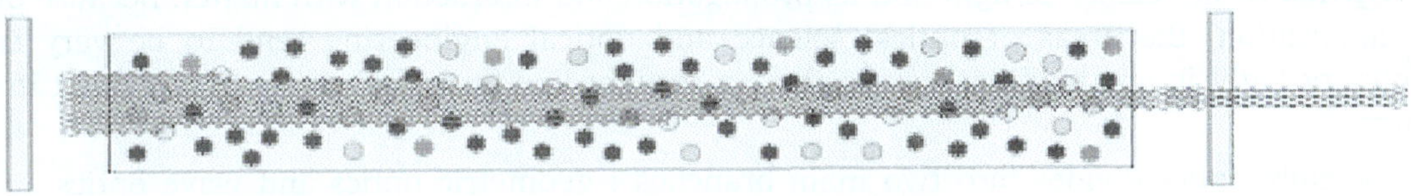

Fig. 8. *Attenuation of light waves under normal equilibrium condition.*

The basic working principle of laser lights is optical amplification. To achieve this a material medium has to be kept in a state of ***population inversion,*** in which the number of atoms in the upper energy level is greater than that in the lower energy level. In this state, when a light beam of appropriate frequency propagates through the atomic collection, there would be more stimulated emission than absorption thus leading to optical amplification. Because in this process the energy levels of the specific atomic system are involved, therefore active medium can only provide for amplification in a band of frequencies around certain specific frequencies.

Fig. 9. *Full stimulated emission, coherent laser beam generated.*

The Pump :

An external source of energy is required to create a state of population inversion, because we need to take the atoms from the lower energy state to the upper state. This external source is called the ***pump.*** The pump could be in the form of optical energy such as in solid-state laser and electrical energy such as in semiconductor lasers or gas lasers. The energy from the pump is essentially absorbed by the medium and then released a part of it in the form of the output laser beam. The rest of the energy supplied by the pump is converted in other forms such as heat etc.

The optical resonator :

Using the same principle of electronic oscillator, an optical amplifier can be converted into an optical oscillator by feeding back a part of the output optical energy into the system. Optical feedback can be easily achieved by using ***reflectors.*** For optical feedback the active medium is placed between a pair of mirrors. One of the mirrors is made fully reflecting and other partially reflecting to allow the laser beam to escape from the cavity and form the output beam.

The lasing action can continue indefinitely if the amplification provided by the medium in one round trip can overcome the loss of the light wave due to partial reflection and other loss mechanisms such as scattering etc., only then we will have the laser emitting a continuous beam of light. Of course, the pump needs to supply energy continuously to maintain the population inversion in the medium to amplify the light wave.

If the gain is higher than the loss, then the intensity of the beam must continuously increase and similarly if the gain is less than the loss, the intensity of the beam must continuously decrease. The light wave forms a standing wave pattern within the ***resonator.*** The number of wavelength within the resonator is very large because the wavelength of light is very small compared to the length of the resonator. Hence the source produces energetic laser light.

In studying optics we want to know what our observations tell us, not about the objects, but about light itself.

Optics is the science of light and its propagation and interaction with matter. Because of the wide application, the areas of optical science and optical engineering tend to be very cross-disciplinary. Light in any medium travels along a straight line from the source to any point in that medium.

Basically optics divides into two main branches : geometric optics and wave optics.

Geometric optics, or ray optics :

It describes light propagation in terms of "rays". Rays are bent at the interface between two media which are not similar and may be curved in a medium in which the refractive index is a function of position. In geometric optics the "ray" is an abstract object which is perpendicular to the wave fronts of the actual optical waves. Geometric optics explains rules for propagating these rays through an optical system.

A **ray** is a line drawn from one wave crest to another which intersects each crest at right angles, as shown in Fig. 10. For light waves, the rays always point in the direction of the motion. Rays therefore provide a useful representation for describing the motion of light waves.

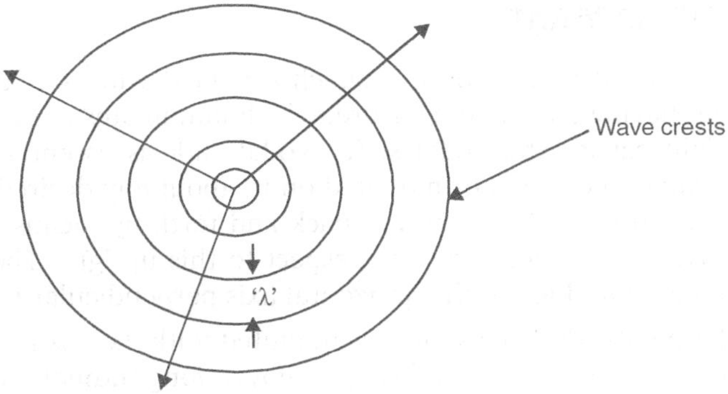

Relationship of rays to wave crests

Fig. 10

EXPERIMENT—14

AIM

To determine the focal length of the combination of two thin convergent lenses separated by a distance, with the help of a nodal slide assembly and to verify the formula

$$\frac{1}{F} = \frac{1}{f_1} + \frac{1}{f_2} - \frac{x}{f_1 f_2}$$

where F = Focal length of the combination of lenses,

f_1 and f_2 = Individual focal lengths of the given lenses,

x = Separation of the two lenses when mount together in optical bench.

APPARATUS REQUIRED

Nodal slide assembly (optical bench, plane mirror, cross slit and lamp) and two convex lenses.

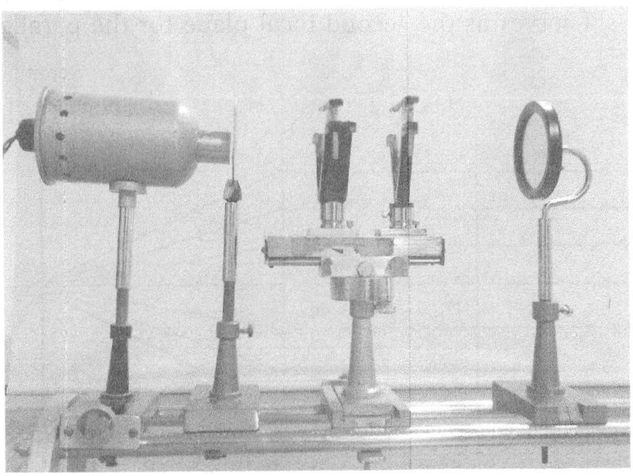

Fig. 14.1. *Nodal slide assembly.*

DESCRIPTION OF THE APPARATUS

The nodal slide assembly consists of an optical bench provided with four uprights carries a bulb placed in a metallic cover having a circular aperture, which illuminates the cross slit in the adjacent upright. The third upright carries the nodal slide. Nodal slide is essentially a horizontal metal support, which permits rotation of the lens mounted on it, about any desired point on its axis. The metallic support can be fixed or it can be moved back and forth by means of a screw so that the relative position of the two lenses can vary with respect to this upright. The forth upright carries plane mirror, which can be rotated about the horizontal axis perpendicular to the bed of the bench.

To locate the nodal points, the lens system is mounted with its axis horizontal on a carriage, which can be rotated about a vertical axis and can be moved longitudinally with the help of a rack and pinion arrangement. This apparatus is known as a nodal slide.

PRINCIPLE

The focal length of a lens system is the distance between its principal point and the corresponding focal point. The principal points coincide with the corresponding nodal points when the media are the same on both the sides of the system (here, air). Thus the focal length of the system can be determined by locating a nodal point and the corresponding focal point.

The second nodal point can be located by using the fact that in case of parallel beam of light incident on a convergent lens system, thus forming an image on a screen in its second focal plane, the image does not shift laterally when the system is rotated about a vertical axis passing through its second nodal point.

The distance between the uprights carrying the screen (or cross slit) and the nodal slide (which gives the position of the axis of rotation) will, therefore, give the focal length of the lens system.

THEORY

Light from a source is sent through a slit that is adjusted to lie at the secondary focal point of the lens (It is an image point on the principle axis of an optical system for which the object point lies at infinity). In this position light diverging from the cross-slit emerges as a parallel beam of light after passing through the lens. This parallel beam of light is reflected as a parallel beam from the plane-mirror and brought to focus on the plane of the cross-slit by the lens. In other words, the screen having the cross-slit serves as the second focal plane for the parallel beam of light coming

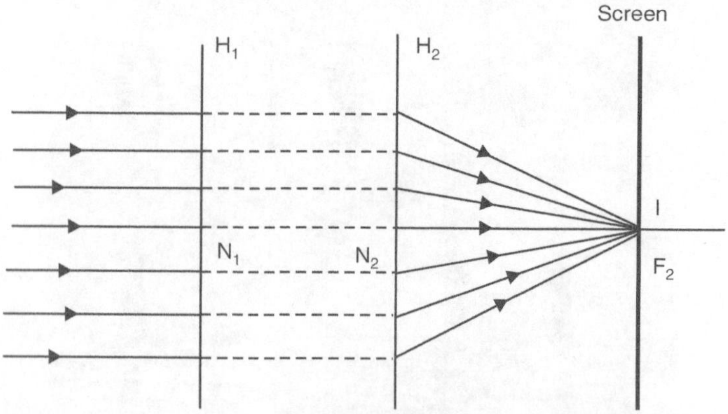

Fig. 14.2

from the plane-mirror. After refraction the emergent beam converges to the second focus F_2 and a real image I is formed on the screen. In the Fig. 14.2 rays along the axis passes through N_1 and N_2, which are the nodal points of the optical system (These are a pair of conjugate points having unit positive angular magnification). The nodal slide carrying the lens system is now rotated back and forth and the lens repeatedly shifted, until the rotation produces no motion of the image I.

In the Fig. 14.3, the lens system is rotated about a transverse axis through O which lies between N_2 and F_2 . N_1 and N_2 take new positions N_1' and N_2' . A ray incident at N_1' and directed towards the first nodal point N_1 takes the path $N_2' I_1$ is such that $N_2' I_1$ is parallel to the incident ray. Since the incident beam is parallel, the image must lie on the second focal plane. Consequently I_1, the point of intersection of the ray $N_2' I_1$ with the focal plane, is the new position of the image (Fig. 14.3). Thus when the axis of rotation lies between N_2 and F_2, a slight rotation of the system in any direction moves the image in the opposite direction.

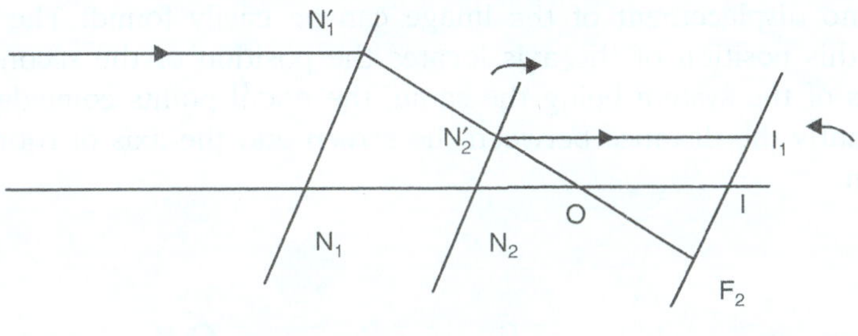

Fig. 14.3

Next suppose the system is rotated by a small angle about a transverse axis through N_2. Then N_1 takes its new position N_1', while N_2 remains fixed. Consequently the position of the parallel emergent ray N_2I is unchanged and the image remains stationary at I (Fig. 14.4). Thus rotation of a lens about its secondary nodal point shifts the refracted rays but not the image.

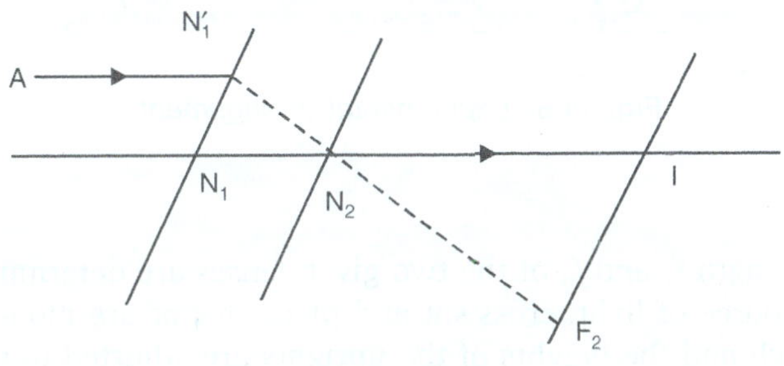

Fig. 14.4

If the axis of rotation passes through a point O_2 which is away from N_2, then a slight rotation of the system displaces both N_1 and N_2 . A ray AN_1' directed towards N_1 takes the path $N_2'I_2$ and the new position of the image is I_2 (Fig. 14.5).

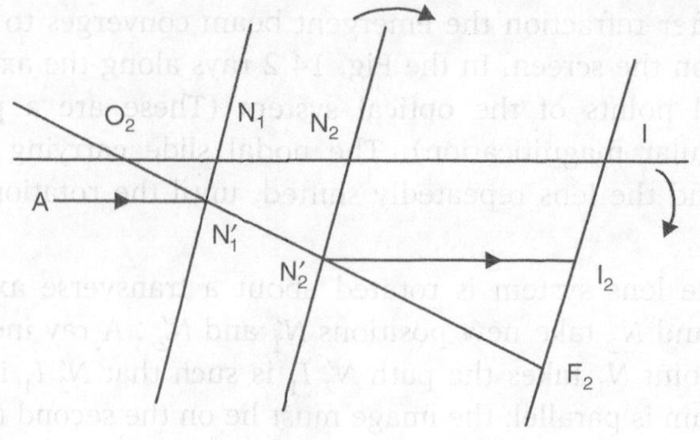

Fig. 14.5

Thus, when the axis of rotation is away from N_2, a slight rotation of the system in any direction displaces the image in the same direction. In this way the position of the axis of rotation, for which there is no displacement of the image can be easily found. The intersection of the principal axis with this position of the axis locates the position of the second nodal point. The media on both sides of the system being the same, the nodal points coincide with the principal points and consequently the distance between the screen and the axis of rotation gives the focal length of the system.

DIAGRAM

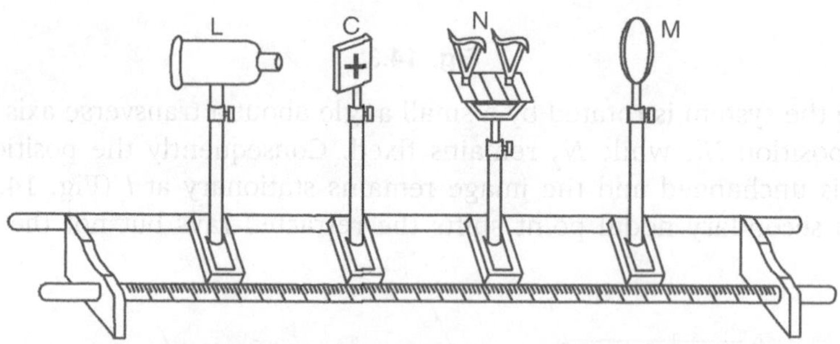

Fig. 14.6. *Experimental arrangement.*

PROCEDURE

1. First the focal length f_1 and f_2 of the two given lenses are determined. For this one of the lenses L_1, the source of light, cross slit and plane mirror are mounted on the uprights of the optical bench and the heights of the uprights are adjusted in such a manner that the line joining the center of each part is parallel to the bed of the bench.
2. Illuminate the cross slit and adjust the plane of the mirror to get the image of cross slit very near to it. The image may be blurred but the well-defined image is found by moving the upright of nodal slide away or towards the cross slit.
3. The slide is rotated slightly about the vertical axis and lateral shift of the image is observed. If there is any shift, the position of the axis of rotation with respect to the lens is slightly changed by moving the nodal slide on the upright by means of the screw provided for this purpose. The sharpness of the image is disturbed. The image is refocused by moving the upright (carrying the nodal slide) on the optical bench. Lateral shift of the

image is again observed. The same process is repeated till the image of the slit is in sharp focus and does not show any lateral shift when the nodal slide is slightly rotated about its vertical axis. The distance between the plane of the cross slit and the axis of rotation now gives the focal length of the lens.

4. The lens is rotated through 180° and the whole process is repeated. The mean of the two distances, thus obtained, will give the exact focal length "f_1" of the lens.
5. The first lens is removed and the second lens is mounted on the nodal slide. Its focal length "f_2" is determined in the same manner as described.
6. To determine the focal length of the combination, the two lenses are mounted on the nodal slide at some distance apart (the lenses are being placed equidistance and on opposite sides of the axis of rotation). By adjusting the inclination of the plane mirror and the position of the nodal slide the image of the cross slit is made to lie on the side of the slit itself. The shift in the image due to a slight rotation of the nodal slide is observed. If there is any lateral shift, with the simultaneous focusing of image a suitable position of the nodal slide is determined for which no lateral shift of the image occurs due to a slight rotation of the nodal slide. The distance between the plane of the screen and the axis of the rotation of the nodal slide now gives the focal length of the combination.
7. Different sets of reading are to be taken by turning the faces of the lens through 180° and interchanging the position of the component lenses.
8. The experiment is repeated for different values of x, the distance between the lenses (say 4, 6, 8 cm).
9. The focal length of the combination is also obtained theoretically for each value of x by the formula

$$\frac{1}{F} = \frac{1}{f_1} + \frac{1}{f_2} - \frac{x}{f_1 f_2}$$

$$F = \frac{f_1 f_2}{f_1 + f_2 - x}$$

It will be found that the experimental and theoretical values of the focal length of the combination for given separation agree fairly well thus verifying the truth of the formula.

OBSERVATIONS

Table 14.1. Determination of focal length of the first lens.

S.No.	***Light incident on***	***Position of cross-slits (a) (cm)***	***Position of the axis of rotation of the nodal-slide (b) (cm)***	***Focal length of the lens $f_1 = (b \sim a)$ (cm)***
1.	One face			
2.	Other face			
			Mean focal length	

Table 14.2. Determination of focal length of the second lens.

S.No.	Light incident on	Position of cross-slits (a) (cm)	Position of the axis of rotation of the nodal slide (b) (cm)	Focal length of the lens $f_2 = (b \sim a)$ (cm)
1.	One face			
2.	Other face			
			Mean focal length	

Table 14.3. Determination of focal length of the combination.

S. No.	Distance between lenses (x) (cm)	Light incident on	Position of cross-slits (A) (cm)	Position of the axis of rotation of the nodal slide (B) (cm)	Focal length of the lens, F = (B ~ A) (cm)
1.		One lens Other lens			
2.		One lens Other lens			
3.		One lens Other lens			

CALCULATION

$$\frac{1}{F} = \frac{1}{f_1} + \frac{1}{f_2} - \frac{x}{f_1 f_2}$$

$$F = \frac{f_1 f_2}{f_1 + f_2 - x}$$

f_1 = ... cm

f_2 = ... cm

1. For distance x = ... cm F = ... cm
2. For distance x = ... cm F = ... cm
3. For distance x = ... cm F = ... cm

By formula :

F = ... cm

By experiment :

F = ... cm.

Similarly calculate for (2) and (3) reading.

RESULT

1. Focal length of the combination of the two given lenses at a distance of

   (*i*) …….. cm apart = ……… cm.

   (*ii*) …….. cm apart = ……… cm.

   (*iii*) …….. cm apart = ……… cm.

2. The calculated and the experimental values of the focal length of the combination being very approximately equal, the formula.

$$\frac{1}{F} = \frac{1}{f_1} + \frac{1}{f_2} - \frac{x}{f_1 f_2} \text{ is verified.}$$

PRECAUTIONS

1. The cross-slits must be properly and intensely illuminated by the light coming from the lamp, which may preferably be a frosted one.
2. The lens or the lens combination should be mounted such that its principal axis passes through the point of intersection of the cross-slits.
3. Sometimes a bright image is formed on the metal plate screen by reflection from the surface of the lens. This image should not be confused with the desired image which is formed by the light which starts from the cross-slits, passes through the lens or lens system, falls on the plane mirror and is reflected back from it through the lens or lens system on the metal plate screen. If the plane mirror is slightly turned the genuine image will shift whereas the false image will remain fixed. This is the test, which can be applied to get the genuine image for the experiment.
4. The rotation of the nodal slide about the vertical axis while testing its coincidence with the nodal point should not exceed 5° or so.
5. When the axis of rotation of the nodal slide passes through the nodal point, the image of the cross-slits will not shift on rotating the nodal slide. If the axis of rotation is nearing the nodal point and the image shifts, say to the left, then when the axis of rotation crosses the nodal point, the image will shift to the right on rotating the nodal slide in the same direction. Thus the desired position can be accurately set.
6. While determining the focal length of a lens, the reading of the axis of rotation on the linear base scale of the nodal slide should be properly recorded and applied.

CONCLUSION

VIVAVOCE

1. What is a lens ?
2. How does a lens refract the rays ?
3. What is focal plane of the lens ? What is principal axis and optical centre ?
4. How is image formed by a convex or concave lens ?
5. What are cardinal points ?
6. What are principal points and principal plane ?
7. What are focal points ?
8. Why do you call it nodal slide ?
9. What are nodal points ?
10. What is the principle of the method ?
11. How will you make sure the image used by you is the desired one ?
12. Will the position of the desired image be affected in any way by altering the position of the plane mirror ?
13. Is it necessary to rotate the combination by 180° ?
14. Which formula are you using in this experiment ?
15. How much a ray suffers deviation in a lens ?

❑❑❑

WAVE OPTICS

In our optics lab we have described the properties of light simply in terms of rays, using the laws of reflection and refraction. This description works well for many phenomenons, including image formation with lenses and mirrors, but for other circumstances it fails. In this lab we examine some phenomenon, which clearly require a different description. This alternative description is called wave optics, to distinguish it from the ray description, which is called geometric optics.

Physical optics or wave optics builds on Huygen's principle and explains the propagation of wavefronts through optical systems, including both the amplitude and the phase of the wave. This technique, can account for diffraction, interference, and polarization effects, as well as aberations and other complex effects. We first introduce some terminology associated with waves.

- The **wavelength (λ)** is the distance between neighboring crests or troughs.
- The **speed (v)** is the rate at which the crests (or troughs) move forward.
- The **period (T)** is the time that elapses between passing crests (or troughs). The period can be expressed in terms of the speed and wavelength $T = \lambda/v$
- The **frequency (f)** is the number of crests (or troughs) that pass by per unit time. It is equal to the inverse of the period.

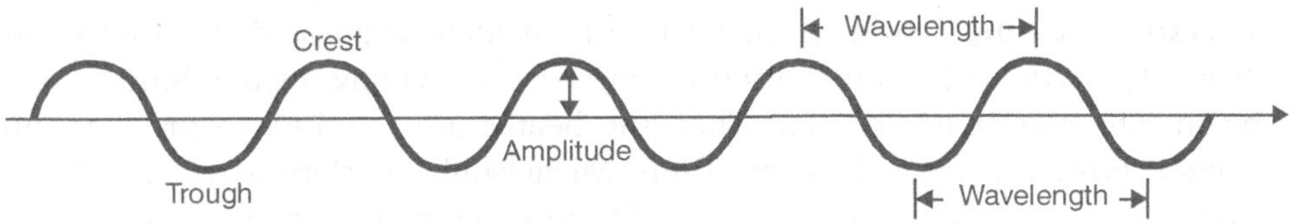

Fig. 1

Using the expression for T above, we get the useful expression.

$$v = f\lambda$$

or Wavelength = Speed of light in vacuum/Frequency.

INTERFERENCE

If the waves have the same sign (are in phase), then the two waves constructively interfere, the net amplitude is large and the light intensity is strong at that point. If they have opposite signs, however, they are out of phase and the two waves destructively interfere; the net amplitude is small and the light intensity is weak. It is these areas of strong and weak intensity, which makeup the interference patterns we will observe in this experiment. Interference can be seen when light from a single source arrives at a point on a viewing screen by more than one path. Because the number of oscillations of the electric field (wavelength) differs for paths of different lengths, the electromagnetic waves can arrive at the viewing screen with a phase difference between their electromagnetic fields. If the electric fields have the same sign then they add constructively and increase the intensity of light, if the electric fields have opposite signs they add destructively and the light intensity decreases.

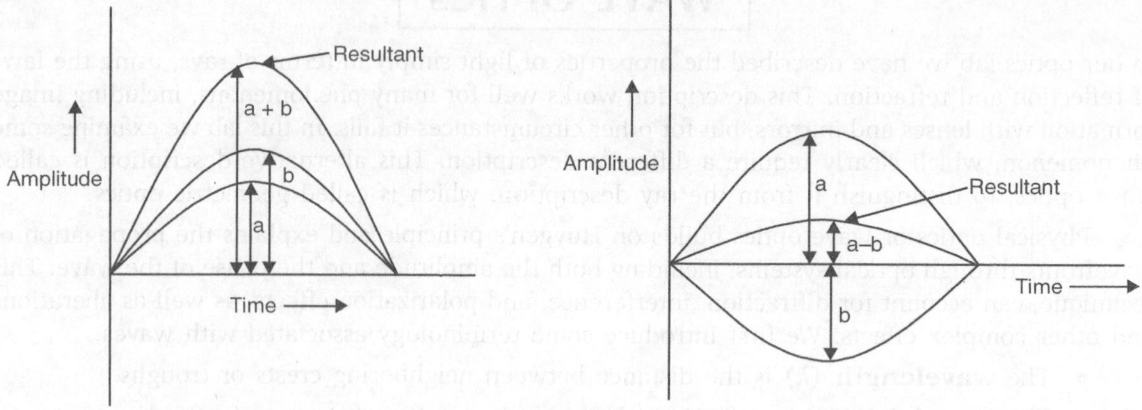

(a) Constructive interference (b) Destructive interference

Fig. 2

Classification of interference phenomenon : The phenomenon of interference of light may be divided into the following two general classes:

(*i*) **Division of amplitude.** In this class the incident beam is divided into two or more beams by partial reflection and refraction, thereby dividing the amplitude of the incident beam into two or more parts. Then the beams are reunited to produce interference fringes. Fringes produced by thin films, Michelson's interferometer and Newton's rings experiment are example of this class. In the experiments of this class broad source of light is used.

(*ii*) **Division of wavefront.** In this type the incident wavefront is divided into two parts by some optical device. Then after they have travelled different optical paths, they are reunited to produce interference fringes. Fresenel's bi-prism experiment, Fresnel's double mirror experiment, Loyd's single mirror experiment are examples of this class. In every experiment of this class a narrow slit illuminated with a monochromatic light should be used as a primary source.

EXPERIMENT—15

AIM

To determine the radius of curvature of a Plano-convex lens by Newton's rings method.

APPARATUS REQUIRED

A Plano-convex lens of large radius of curvature, optical arrangement for Newton's ring (optically plane glass plate and a Plano-convex lens), a plane glass plate inclined at an angle of 45°, a travelling microscope and sodium lamp provided with cover having a broad slit cut into it.

Fig. 15.1. *Newton's ring apparatus.*

DESCRIPTION OF THE APPARATUS

Newton's ring assembly consists of Plano-convex lens mounted on an optically plane glass plate. Fig. 15.2 shows the convex surface of a lens in contact with a plane glass plate. A thin film of air is formed between the two surfaces. Above it an adjustable plane glass plate is provided to be inclined at 45° with respect to the vertical plane. When we view the setup using a travelling microscope from the top of the glass plate with monochromatic light, the circular interference fringes are observed. These were studied by Newton and are called Newton's rings. Sodium vapor lamp has been used as the monochromatic (5893 Å) and broad light source. Travelling microscope can move along perpendicular direction. Horizontal measurement scale is fitted with fine and coarse movement screw. Cross-wire in the field of view is for ring's diameter measurement.

PRINCIPLE

With the help of this apparatus, the wave nature of light is confirmed. Production of Newton's rings is due to interference taking place in a film of varying thickness *i.e.,* wedge-shaped film. This type of film can be formed when the convex surface of a long focus Plano-convex lens (of large radius of curvature) is placed in contact with a plane glass plate. Since the curved surface of the lens is spherical the thickness of the film is the same for all points on a circle concentric with the point of contact, and it slowly increases continuously as we proceed away from the point of contact. The film is illuminated with monochromatic light (because if white light is used, then each separate color will give rise to its own complete set of rings. These set of rings will overlap each other with the result that except for a few colored rings near the point of contact, the rest of the region will appear white) and viewed by reflected light in a travelling microscope. A system of concentric circular rings, alternately dark and bright, is observed. The contact point is black by reflected light and bright by transmitted light. These circular fringes are called Newton's rings. They are fringes of equal thickness. The rings are due to interference of light waves reflected from the upper surface (glass-air boundary) and the lower surface (air-glass boundary) of the air film enclosed between the lens and glass plate.

THEORY

When the system is illuminated with light from a monochromatic source, part of the light is reflected from the top of the air film, say, from point B (Fig. 15.2). The remainder of light passes through the air film. BJ represents the thickness t of the air film at a distance r form the center O. The two reflected rays (reflected from B and J) therefore have a net path difference of $2t$ where $t = BJ$. The same path difference is obtained at all points around O, which are at distance $OJ = r$ from O. As the reflected rays are derived from the same incident ray, the two rays are coherent and when they overlap on each other produce interference. Further the two reflected rays are obtained through a division of amplitude of the incident ray. Therefore this kind of interference occurring in the thin film of air belongs to the class of interference through the division of amplitude.

In addition to the geometric path difference $2t$ between the rays reflected at B and J there arises a phase difference 180° for the wave reflected at J. These phase difference corresponds to a path difference of $\lambda/2$ (If the medium beyond the interface has a lower refractive index there is no phase change in the reflected wave due to the reflection. But if the medium beyond the interface has a higher refractive index a phase change of π radian equivalent to the path increase of $\lambda/2$ occurs in the reflected wave, where λ is the wavelength in the rarer medium. The wave transmitted through the interface does not undergo a phase change in either case.) Therefore the optical path difference Δ will be $(2t + \lambda/2) = n\lambda$. In this case the two rays will be in phase and their crests coincide with each other. It causes an increase in the intensity of the reflected light.

On other hand if it satisfies the condition that

$$\Delta = (2t + \lambda/2) = (2n + 1)\lambda/2$$

where n is the order of the ring. The rays will be out of phase and the crest of one wave falls on the trough of the second wave leading to the cancellation of the intensity of the light. Hence darkness is caused.

The condition of interference will be the same for all points wherever the thickness of the film is constant. It is seen above that the thickness of the film is constant along circles around O. Hence the bright and dark regions take the shape of rings as shown in Fig. 15.2.

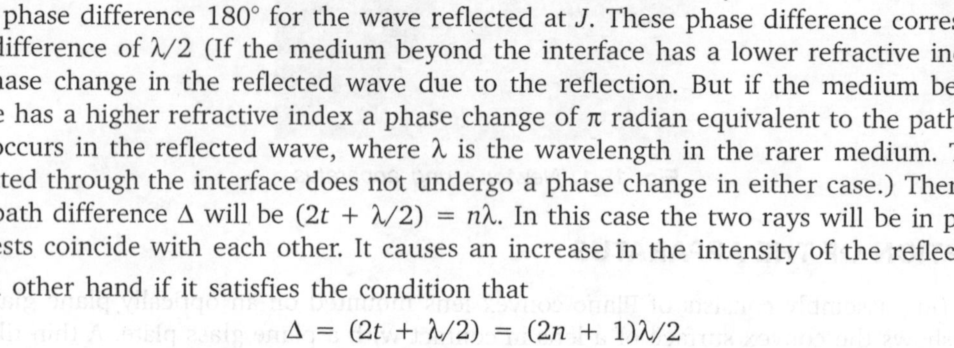

Fig. 15.2. *Ray diagram and pattern.*

At the center O the film thickness t is zero. But because of the phase change that occurs for ray (2) on reflection at the plane glass plate, the two rays will be out of phase by 180°. The crest of ray (1) therefore falls on the trough of ray (2). They cancel each other and give darkness as a result a dark spot is obtained at the center of the fringe system. Because the contact of the lens and the glass plate occur over small area, we get a dark spot of considerable size. It is derived that the radius of a dark ring is related to the radius of curvature of the lens as follows.

$$r_n = \sqrt{nR\lambda} \quad ...(1)$$

for a bright ring

$$r_n^2 = \frac{(2n-1)}{2} R\lambda \quad ...(2)$$

It is seen from equation (1) that the radii of the dark rings are proportional to the square root of natural number. The radii of bright rings are seen to be proportional to the square root of odd natural numbers. Hence as the order of the ring increases, the radius does not increase in the same proportion and the rings get closer and closer.

From (1) we can write for the diameter of nth dark ring as,

$$D_n^2 = 4nR\lambda \quad ...(3)$$

For the $(n + p)$th dark ring

$$D_{(n+p)}^2 = 4(n + p)R\lambda \quad ...(4)$$

From equations (3) and (4), we get

$$R = \frac{D_{(n+p)}^2 - D_n^2}{4p\lambda}.$$

FORMULA USED

$$R = \frac{D_{(n+p)}^2 - D_n^2}{4p\lambda}$$

where, $D_{(n+p)}$ = Diameter of the $(n + p)$th ring, where p is an integer.

D_n = Diameter of the nth ring.

p = An integral number that is the difference in the order of two rings considered.

R = Radius of curvature of the curved surface of the plano-convex lens.

λ = Wavelength of sodium light (λ is known).

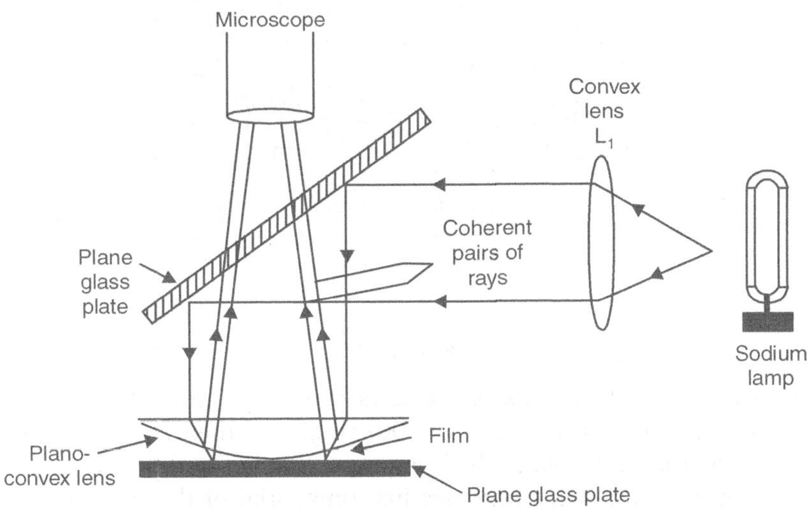

Fig. 15.3. *Experimental arrangement.*

PROCEDURE

1. The surface of both of glass plate and the plano-convex lens should be first cleaned smoothly with handkerchief.
2. A glass sheet inclined at 45° to the beam and positioned vertically above the Newton's ring set up will illuminate the set up properly and provides maximum brightness.
3. In order to focus the microscope, the eyepiece is first focused on the cross-wires till they are clearly seen in the field of view. One of the cross-wire is made horizontal. The microscope is then focused on a cross mark (+) made on a piece of paper placed above the blackened glass plate. The lens has not been put as yet.
4. The paper is removed and the plano-convex lens is put in its place with curved surface in contact with the plate. The microscope is slightly adjusted so that the rings are clearly seen in the field of view. If the rings obtained are not circular then either the glass plate or lens or both are irregular. To obtain perfectly circular rings, lens is shifted to different positions on the glass plate. Sometimes the lens and the plate are not perfect on account of the presence of dust particles but this will not affect the experiment.
5. By moving the blackened glass plate (along with the lens) or the microscope this way or that way the horizontal cross-wire is made to pass through the center of the ring system parallel to the scale so that the motion of this cross-wire may be along the diameter of the ring system.
6. The Newton's ring set up is examined in reflected light to see whether interference fringes are formed or not and whether they are circular or not. Adjustments are to be done till satisfactory fringe system of perfect circular shape with a dark spot at the center is obtained.
7. The least count of the microscope is evaluated and recorded.

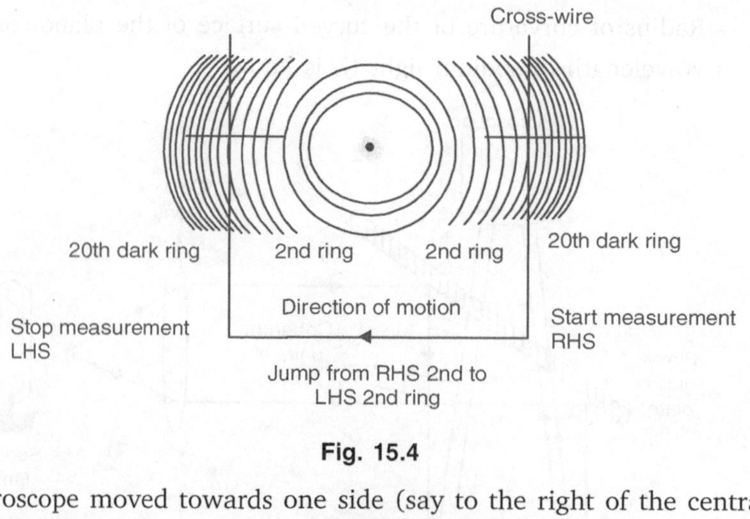

Fig. 15.4

8. The microscope moved towards one side (say to the right of the central spot of certain numbers of dark and bright rings, say 20th ring) and then allowed to move towards the left. As soon as the other wire which is perpendicular to the direction of motion of the microscope of the cross-wires becomes just tangential of the outer edge of 20th dark or bright ring. The reading of microscope screw is noted. The motion of microscope towards

the right is continued slowly by the circular screw and the readings of the screw are recorded every time the cross-wire becomes tangential to the outer edge of the every alternate dark or bright rings (*e.g.,* 20th,18th,16th, etc.)

On reaching the central spot portion, the cross-wires are moved further to the other side of it and the readings of the screw recorded every time the cross-wire becomes tangential to the inner edge of the alternate dark or bright rings and 4th, 5th, etc., till 20th dark or bright ring is reached. Difference between the microscope readings on the outer edge on one side and inner edge on another side of a ring will give its mean diameter.

9. Tabulate your readings as shown in Table 15.1 and compute D_n and D_n^2.

OBSERVATION TABLE

In travelling microscope

Value of one division of main scale = ... cm.

No. of division on the vernier scale = ...

Therefore,

$$\text{Least count of the travelling microscope} = \frac{\text{Value of one division on main scale}}{\text{Total number of divisions on vernier scale}}.$$

Table 15.1. Table for measuring diameters of rings.

No. of rings	***Microscope-reading when the cross-wires is tangential to the***						***Diameter of the ring*** $D_n = (a \sim b)$ ***(cm)***	D_n^2 ***(cm²)***
	Outer edge (R.H.S.) ↓ (cm) (a)			***Inner edge (L.H.S.) ↑ (cm) (b)***				
	M.S.R. (cm)	***V.S.R. (cm)***	***Total (cm)***	***M.S.R. (cm)***	***V.S.R. (cm)***	***Total (cm)***		
20								
18								
16								
14								
12								
10								
8								
6								
4								
2								

GRAPH

Plot a graph between number of ring and square of its diameter. A straight line graph is obtained.

The slop of the graph gives the value of $\dfrac{\left(D^2_{(n+p)} - D^2_n\right)}{p}$

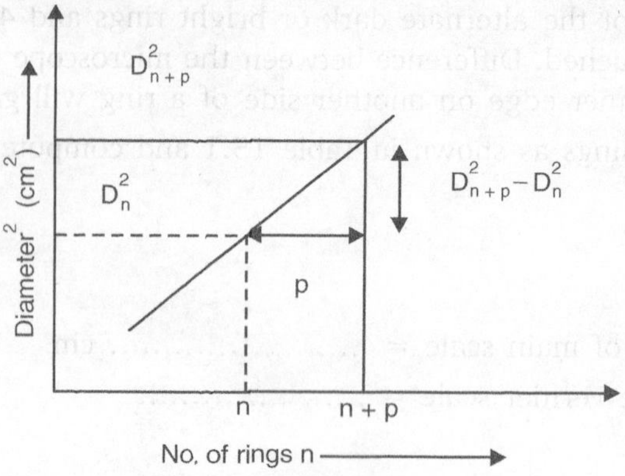

Fig. 15.5

CALCULATION

Knowing the value of the parameters used in the formula the radius of curvature can be calculated with the formula :

$$R = \frac{D^2_{(n+p)} - D^2_n}{4p\lambda}$$

PRECAUTIONS

1. The lens and the glass should be neat and clean.
2. The convex lens should have a large radius of curvature.
3. The slow motion screw of the microscope should be moved in the same direction to avoid backlash error.
4. An extended source of light should be used.

CONCLUSION

VIVAVOCE

1. What do you mean by interference of light ?
2. What does interference of light prove about the nature of light ?
3. What are the necessary conditions for getting the interference of light ?
4. What is the principle of superposition in wave motion ?
5. What are coherent sources ? How do you produce coherent sources ?
6. Can two independent monochromatic sources be coherent ?
7. Name the two classes of interference.
8. Why are the Newton's rings circular ?
9. Why is the central spot dark ?
10. How can we form Newton's rings with white center ?
11. What is the effect of :
    (*a*) Increasing the curvature of the lens on the diameter of the rings ?
    (*b*) Introducing water between the convex lens and the glass plate ?
    (*c*) Changing the Na-light to the mercury-lamp ?
12. What is the use of interference phenomenon ? Can you determine μ of a liquid by Newton's rings ?
13. Why a glass plate is placed at an angle 45° ?
14. Why the readings are taken by coinciding the cross-wire with dark rings ?
15. What do you mean by constructive and destructive interference of light ?
16. Give reason for the cause of an additional phase change of π-radian in Newton's ring formation for the reflected rays of light.
17. Which part of the plano-convex lens is placed over blackened glass plate ?
18. Due to which optical phenomenon of light Newton's rings are observed ?
19. What is the use of 45° inclined glass plate ?
20. What are the optical path differences for observing maxima and minima of light intensity ? Also give expressions for the radius of bright and dark Newton's rings.

❑❑❑

EXPERIMENT—16

AIM

To determine the fringe width (The spacing between two adjacent bright or dark fringes) of an interference pattern with the help of Fresnel's bi-prism.

APPARATUS REQUIRED

An optical bench with four vertical stands, bi-prism, sodium lamp (monochromatic source) eye-piece provided with micrometer screw, slit, bi-convex lens.

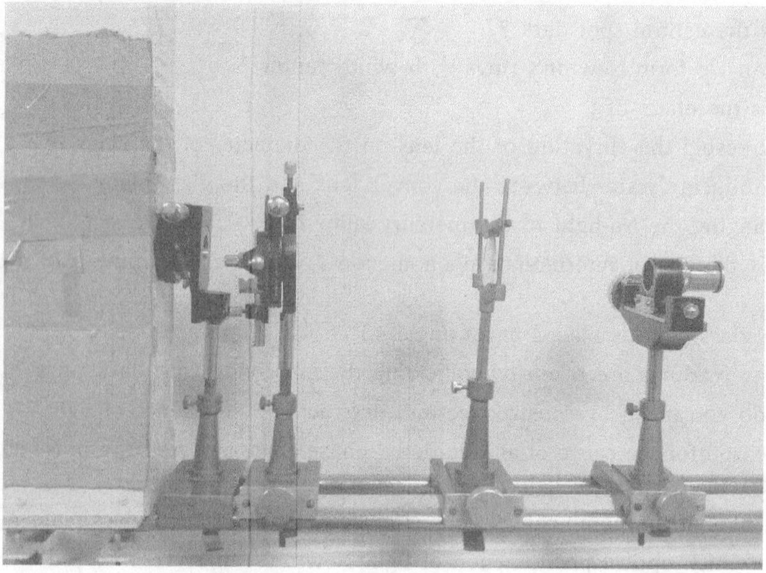

Fig. 16.1. *Fresnel's bi-prism setup.*

DESCRIPTION OF THE APPARATUS

Fresnel's bi-prism is a single prism, made of two right angle prism whose refracting angles are nearly 30' placed base to base. This prism has three angles 30', 30' and 179°. In practice a very thin glass plate is grounded, polished and given a shape such that it behaves like a bi-prism (Fig. 16.2).

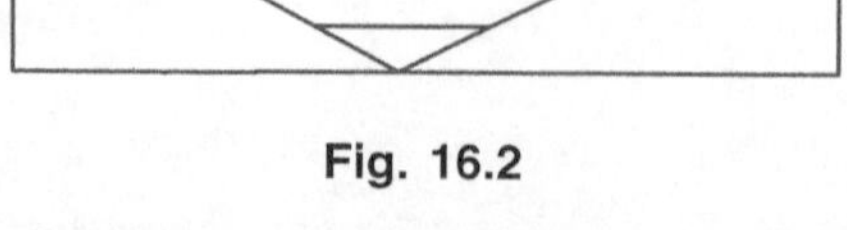

Fig. 16.2

Optical bench consists of a heavy base made of two iron rods, each nearly 1.5 m long on four levelling screws. On the front face of one rod a scale is graduated in mm. The bench is provided with four vertical stands respectively for slit, bi-prism, convex lens and micrometer eyepiece.

Each vertical stand,

1. Can be moved along the length of optical bench,
2. Can be moved along the width of the optical bench,
3. Can be rotated about the vertical axis of stand, and
4. Can be moved up and down perpendicular to the optical bench.

Apart from this, the slit and bi-prism stand can also be rotated in their vertical planes with the help of tangential screws provided on them. Each stand can be clamped at any height as desired. The micrometer eyepiece is generally Ramsden's eyepiece provided with cross-wire. This can be moved on the stand along the width of the optical bench with the help of the micrometer screw and its reading can be read with the help of main scale and circular head scale. Light source sodium lamp (monochromatic source) contained in a rectangular black box.

PRINCIPLE

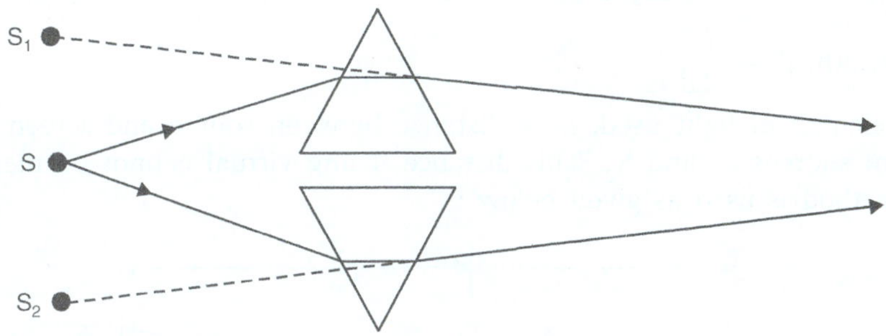

Fig. 16.3

In this arrangement, two coherent sources are produced out of one parent source with the help of a bi-prism. The light from S is refracted from the upper and lower parts of the bi-prism and bends towards the base of the prism and the rays emerging out from these prisms appear to come from S_1 and S_2 (as shown in Fig. 16.3). Two virtual images S_1 and S_2 serve as coherent sources for producing stationary interference pattern. Therefore this kind of interference belongs to the class of interference through the division of wavefront. The distance between any two consecutive bright or dark fringes is same and this distance is called the fringe-width.

THEORY

The working of bi-prism is shown in Fig. 16.4; S is a narrow vertical slit, which is illuminated with a monochromatic light source (sodium light source). It is placed nearly 5 cm from the bi-prism such that it is parallel to the edge of the bi-prism. The screen is illuminated by source S_1 over the region RY and by source S_2 over the region XL. Hence equally spaced alternately bright and dark interference fringes occur near the centre of the region XY (where lights from two sources overlap) and can be observed by a micrometer eyepiece. In this position, both the coherent sources S_1 and S_2 are of equal size and of same intensity so that both the interfering waves are also of nearly equal amplitudes. Although these interference fringes are hyperbolic, but their eccentricity is so high that they appear to be nearly straight.

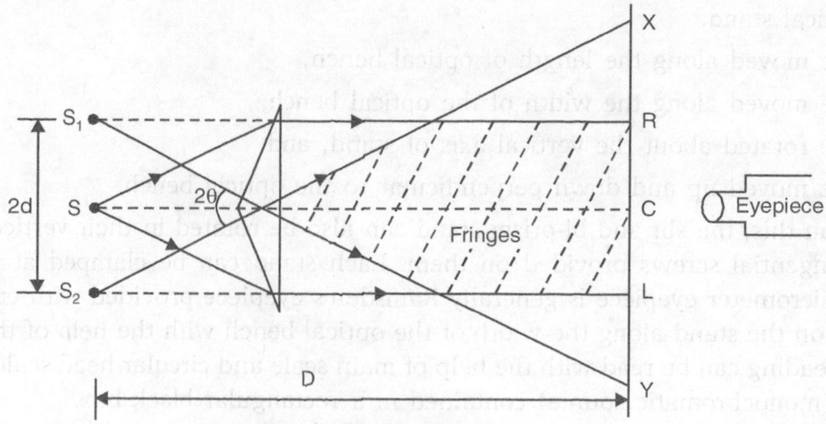

Fig. 16.4

The spacing between two adjacent bright (or dark) fringes is given by the relation

$$\text{Fringe width, } \beta = \frac{D\lambda}{2d} \qquad ...(1)$$

where λ = wavelength of light used, D = distance between source and screen, $2d$ = distance between coherent sources S_1 and S_2. This distance, being virtual cannot be measured directly; hence indirect method is used as given below :

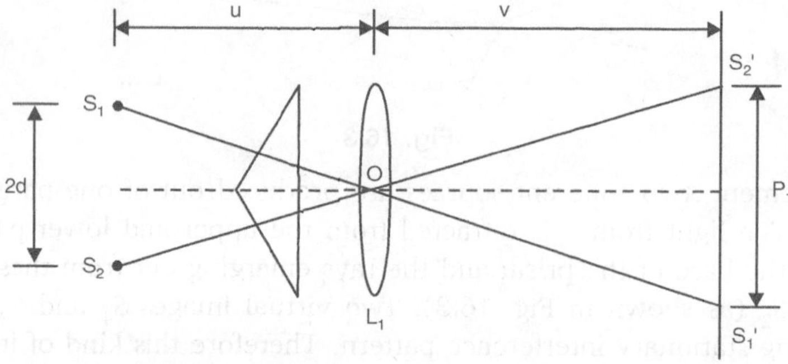

Fig. 16.5

In this method (Fig. 16.5) a suitable convex lens is put in position L_1 on the third vertical stand on the optical bench, in between the bi-prism and the eyepiece between the source S and the eyepiece M. Without disturbing their position, the position of the lens is adjusted till sharp images of S_1 and S_2 are obtained in the field of view of the eyepiece. The two images obtained are S_1' and S_2'. The distance between these images is measured. Let it be P_1. Then from the two similar triangles, we get

$$\frac{S_1S_2}{P_1} = \frac{u}{v}$$

or

$$\frac{2d}{P_1} = \frac{u}{v} \qquad ...(2)$$

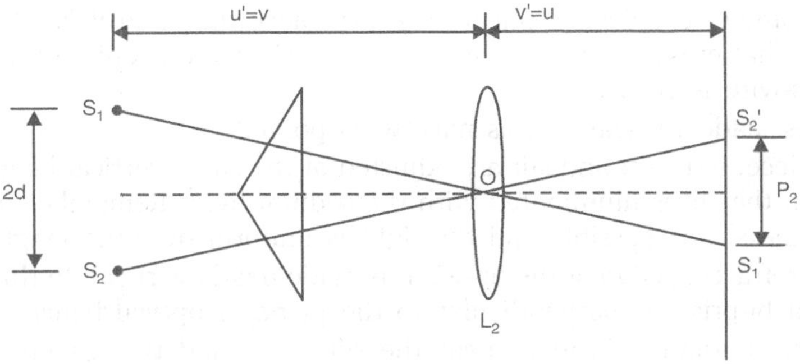

Fig. 16.6

The lens is now moved to the second position L_2 as shown in Fig. 16.6. Again we will get, in sharp focus, a pair of images of the source. Let the distance between these two images be P_2 (In the first case, P_1 is greater than the distance S_1S_2 but in the second case P_2 is as much smaller than S_1S_2). Again from the two similar triangles, we have

$$\frac{S_1S_2}{P_2} = \frac{u'}{v'}$$

Now $u' = v$ and $v' = u$ (because object and image are conjugate)

Therefore $$\frac{2d}{P_2} = \frac{v}{u} \qquad ...(3)$$

Multiplying (2) and (3), we have

$$\frac{2d}{P_1} \times \frac{2d}{P_2} = 1$$

or $$\frac{(2d)^2}{P_1P_2} = 1$$

or $$2d = \sqrt{P_1P_2}$$

Hence equation (1) becomes

Fringe width $$\beta = \frac{D\lambda}{\sqrt{P_1P_2}}$$

FORMULA USED

$$\beta = \frac{D\lambda}{2d}$$

where λ = Wavelength of monochromatic light.

D = Distance between coherent sources and micrometer eyepiece.

$2d = \sqrt{P_1P_2}$ = Virtual separation between sources.

PROCEDURE

1. The base of optical bench is made horizontal with the help of spirit level and the four levelling screws provided at the base of optical bench.

2. The micrometer eyepiece is put into the eyepiece tube and it is adjusted by moving in or out so that cross-wires are distinctly seen. Then the eyepiece tube is rotated so that one cross-wire is vertical.
3. The slit is made vertical and as narrow as possible.
4. The eyepiece, bi-prism and slit are adjusted at the same vertical height and in a straight line. Then the slit is illuminated with the sodium light. Remember that the source is as close to the slit as possible and the light is uniform over the length of optical bench.
5. Now the refracting edge of the bi-prism is made nearly vertical. In this position the plane surface of bi-prism is perpendicular to the plane of optical bench.
6. The bi-prism stand is brought near the slit stand and two coherent sources are seen through the bi-prism. If two coherent sources are not visible, the bi-prism is moved perpendicular to the optical bench till on seeing through the bi-prism along the length of the optical bench, two uniform images of coherent sources are seen.
7. Now the eyepiece stand is brought near the bi-prism stand and moving the eyepiece perpendicular to the optical bench alternate bright and dark fringes are seen in the field of view through the eyepiece. This is the region where the interference fringes are formed due to superposition. Then the bi-prism is rotated in its plane by means of its tangential screw till distinct fringes are seen.

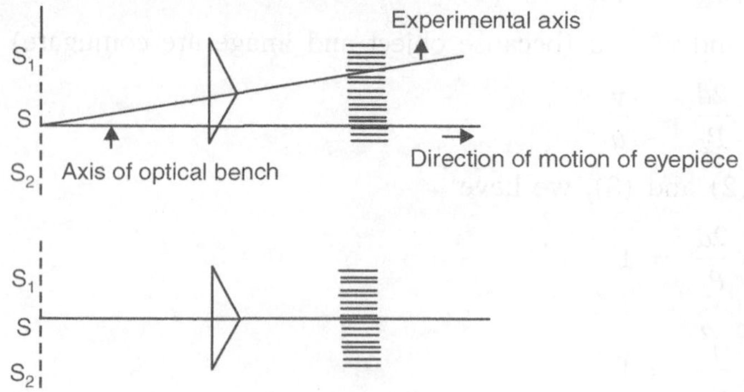

Fig. 16.7

8. Now seeing the fringes through the eyepiece the eyepiece stand is moved away from the bi-prism. In doing so, if the fringes appear to shift towards left or towards right, from the field of view, then it is called lateral shift of the fringes. It happens when the line joining the slit and refracting edge of bi-prism (the experimental axis) is not parallel to the axis of optical bench. To remove it, the eyepiece stand is brought near the bi-prism stand and the cross-wire is made to coincide on a bright fringe. Then the eyepiece stand is slightly moved away and the bi-prism is shifted slightly perpendicular to optical bench in a direction opposite to the direction of displacement of the fringes. Then the lateral shift of fringes is again tested and again the bi-prism is moved in direction opposite to the direction of displacement of fringes. The process is repeated till the lateral shift is completely removed, means on moving the eyepiece upright away from bi-prism, position of central fringe coincides the vertical section of cross-wire and equal deployment of fringes on either side of central fringe are seen.
9. In order to see the images of two coherent sources, convex lens is introduced between bi-prism stand eyepiece stand. Initially distance between the plane of source and plane of screen should be greater than or equal to four times of focal length of lens used.
10. Now move this lens stand from bi-prism to eyepiece and will observe the sharp images of coherent sources in different position of lens.

11. In the first position of lens (when lens is kept near to bi-prism), sharp images of coherent sources are seen in the field of view of eyepiece. Now find out the separation between images *i.e.,* P_1. This can be calculated using micrometer attached in eyepiece.
12. Keep the vertical section of the cross-wire on the one of the images. Take the observation and move it to second image and take observation, so difference in observation will be P_1 (size of image).
13. Now slide the lens from 1st position to eyepiece and observe the clear and sharp images of coherent sources on the eyepiece and find the separation between images *i.e.,* P_2, using micrometer with the same method mentioned in point 12.
14. Now separation between the two coherent sources ($2d$) can be calculated by using P_1 and P_2 by formula $2d = \sqrt{P_1 P_2}$.
15. After getting $2d$, β can be calculated using formula $\beta = \frac{\lambda D}{2d}$.

OBSERVATION TABLE

For determination of fringe width :

$$\text{Pitch} = \frac{\text{Distance travelled on the pitch scale}}{\text{Number of rotations of circular head}} = \ldots\ldots\ldots\ldots\ldots \text{ cm}$$

(If 2 rotations are given to the circular scale, the distance moved is 1 mm, the Pitch $= \frac{1}{2}$ $= 0.5$ mm)

Number of divisions on the circular scale = ...

$$\text{Least count of micrometer screw} = \frac{\text{Pitch}}{\text{Total number of divisions on circular scale}} = \ldots\ldots\ldots\ldots \text{ cm}$$

Table for determination of 2*d* :

Table 16.1. For P_1 (size of image when position of lens is near the bi-prism).

S.No.	Position of image of first illuminated slit			Position of image of second illuminated slit			Size of image $P_1 = b \sim a$
	M.S.R.	C.S.R.	TOTAL 'a' = M.S.R+C.S.R	M.S.R.	C.S.R.	TOTAL 'b' = M.S.R+C.S.R	

Table 16.2. For P_2 (size of image when position of lens is near the eyepiece).

S.No.	Position of image of first illuminated slit			Position of image of second illuminated slit			Size of image $P_2 = b \sim a$
	M.S.R.	C.S.R.	TOTAL 'a' = M.S.R. + C.S.R.	M.S.R.	C.S.R.	TOTAL 'b' = M.S.R. + C.S.R.	

CALCULATION

$$2d = \sqrt{P_1 P_2} = \text{...............} \text{ cm.}$$

$$\beta = \frac{D\lambda}{2d} = \text{...............} \text{ cm.}$$

RESULT

The value of fringe width β = ... cm.

PRECAUTIONS

1. The optical bench must be horizontal.
2. The slit must be as narrow as possible and vertical.
3. The refracting edge of the bi-prism must be parallel to the slit.
4. The line joining the slit and refracting edge of the bi-prism must be parallel to the axis of optical bench otherwise there will be lateral shift in the fringes.
5. To avoid backlash error, the micrometer screw should be moved always in one direction.
6. While measuring the fringe width, vertical cross-wire should coincide with the middle of the fringe.
7. The motion of micrometer screw should be normal to the length of optical bench.

CONCLUSION

VIVAVOCE

1. What is a bi-prism ?
2. How is it made ? Explain fringe-width of a bi-prism.
3. Why are the refracting angles of the two prisms made so small ?
4. Where are the virtual sources situated ?
5. On what factors does fringe width depend ?
6. Why should the slit be very narrow ?
7. What is lateral shift ?
8. Why is it necessary to remove the lateral shift ?
9. How do you measure 2*d* ?
10. Where on the fringe do you adjust the cross-wire in measuring the fringe width ?
11. What do you get if you use a white light source ?
12. What do you mean by a wavefront ?
13. What do you understand by the division of wavefront ?
14. How are the fringes formed ?

❑❑❑

DIFFRACTION

Waves of all types exhibit diffraction and it is more pronounced for waves of larger wavelengths ($v = f\lambda$). We can hear around corners because of diffraction of sound waves. Effects produced by diffraction of light waves are usually very small (due to extreme shortness of the wavelength of light) and hence are observable only on very careful examination. In the case of ordinary obstacles (which are million times greater than the wavelength of light) the amount of bending is vanishingly small with the result that light seems to travel in straight lines without any deviation.

When light falls on obstacles and if the dimensions of an obstacle are comparable with the wavelength of light, it deviates from its rectilinear path and bends round the corners of those obstacles. This bending of light round the corners of an obstacle is called diffraction of light. It was found that diffraction or bending of light round obstacles produced bright and dark fringes known as diffraction bands or fringes. It is important to note that diffraction effects are produced only when the obstacles cut-off a part of the incident wavefront.

DIFFERENCE BETWEEN INTERFERENCE AND DIFFRACTION

The important difference between the two phenomena is that in the case of interference, the resultant intensity at a point is due to the superposition of two wavefronts originating from two coherent sources, but in the case of diffraction, the resultant intensity of light at a point is due to the interference of light from the different parts of the same wavefront. Due to this, generally in interference pattern all maxima are equally bright, minima almost perfectly dark, bands are large in number while in case of diffraction, maxima are of decreasing intensity, minima are not perfectly dark and bands are few in number.

The diffraction phenomenon is actually interference phenomenon caused by the diffracted light. But whereas the true interference is obtained with no limitation whatever on the extent of the wavefront, the diffraction is applied to those cases in which one is concerned with the resultant effect produced by a limited portion of a wavefront.

Both are governed by the same physics of superposition and Huygen's principle.

(a) Diffraction (b) Interference

Fig. 1

Diffraction Phenomenon is Divided into Two General Classes

1. Fresnel diffraction

In this class either the source of light or the screen or both are at finite distances from the obstacle. Further in this case no lenses are used and the wavefronts incident on the diffracting aperture or the obstacle and those leaving it are divergent and not plane; they may be spherical or cylindrical. Therefore the corresponding rays are not parallel.

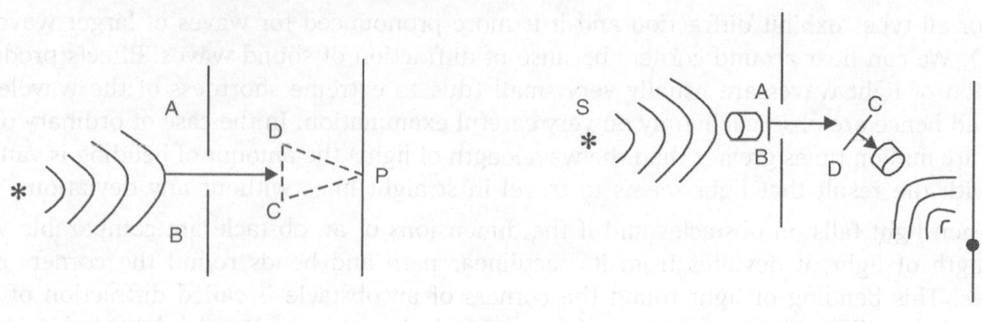

(a) Fresnel class (b) Fraunhoffer class

Fig. 2

2. Fraunhoffer diffraction

The diffraction phenomenon in which the source and the screen are effectively at infinite distances from the diffracting aperture or the obstacle is listed under this class. The wavefronts incident on the diffracting aperture or the obstacle and those leaving it are plane. Therefore the corresponding rays are parallel to one another.

In the laboratory Fraunhoffer diffraction pattern is obtained with the help of spectrometer. Light from an illuminated slit is made parallel by the collimating lens. The diffracting aperture is mounted on the prism table and the diffracted rays are focused in the focal plane of the objective of a telescope. In the Fresnel type of diffraction if the distance between the source and the obstacle and that between the obstacle and the screen are made very large this class of phenomenon merges into the Fraunhoffer class.

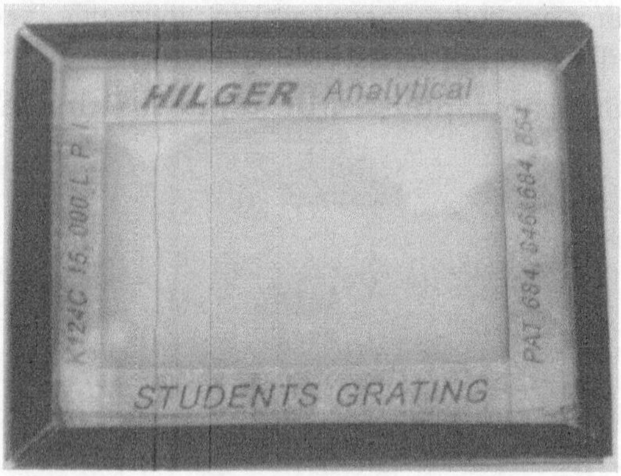

Fig. 3. *Grating.*

A diffraction grating is an optical device. One of the most important and interesting application of diffraction is the Grating. It consists of a very large number of extremely narrow parallel slits of the same width arranged very close together at regular spacing. It is prepared by ruling a large number of fine parallel lines at regular spacing either on a glass plate or on a polished metal by a sharp diamond point. The ruled lines are opaque to light; whereas the clear spaces between the lines transmit light and act as narrow slit. In most of the gratings it is about 15,000 lines per inch. Such gratings are very costly. Therefore cheaper plastic replicas are made.

Let a parallel beam of light (or plane wavefront) of wavelength λ be incident on the grating (Fig. 4). The problem of finding the intensity of light transmitted by the grating involves the principles of interference and diffraction. Since the slits are narrow, each one gives rise to diffracted rays. These diffracted rays then interfere with one another to produce the final pattern that consists of :

(*i*) Widely – spaced diffraction maxima (known as primary maxima such as M_0, M_1 and M_2 etc.)

(*ii*) And a large number of faint subsidiary maxima and minima (such as those of points *P*, *P* etc.) in between the primary maxima.

GRATING ELEMENT

The distance between the centers of two adjacent slits is called the grating element, or the grating spacing. It comprises of one clear space and one ruled space of a grating. It is given by $(a + b)$, where a is the thickness of the opaque ruled line and b is the thickness of transparent un-ruled space between the two ruled lines. Gratings having 15,000 (*N*) lines per inch have been manufactured and are widely used for spectroscopic purposes. Figure represents the section of a grating whose slits are perpendicular to the plane of the paper. Only eight slits have been shown in the figure, although an actual grating contains thousands of them per cm. Obviously, number of slits per unit length of the grating is $N = 1/(a + b)$.

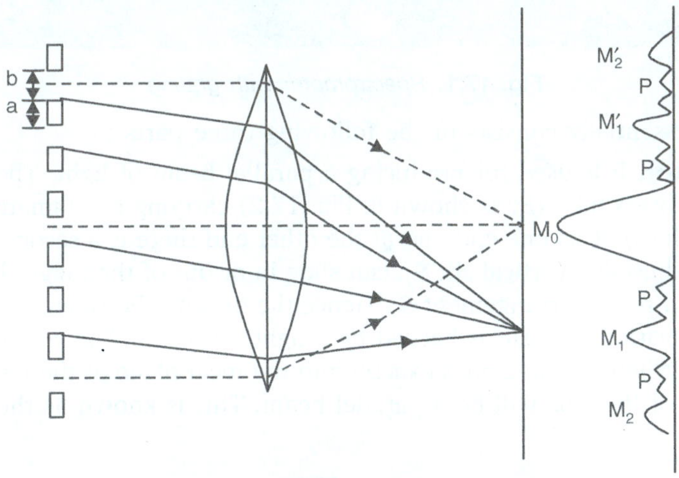

Fig. 4

EXPERIMENT—17

AIM

To determine the wavelength of prominent spectral lines of mercury light using a plane transmission grating.

APPARATUS REQUIRED

Mercury lamp, a spectrometer, a spirit level, a table lamp, a grating, and a reading lens.

DESCRIPTION OF THE APPARATUS

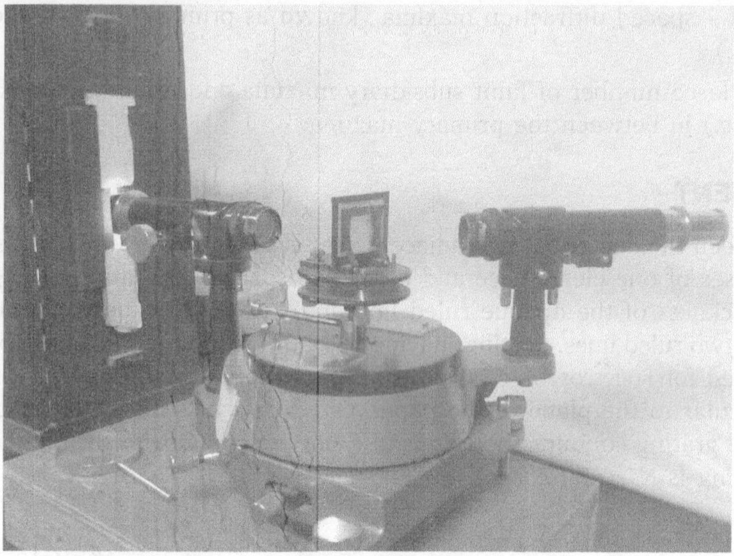

Fig. 17.1. *Spectrometer with grating.*

Spectrometer essentially consists of the following three parts :

1. **A collimator.** It is used for producing a parallel beam of light. The collimator consists of a hollow brass tube (*B*, as shown in Fig. 17.2) carrying a collimating lens (achromatic converging lens) at its one end and at the other end there is a second small tube *A* which carries an adjustable vertical slit S_2, can slide in or out of the bigger brass tube by means of a rack and pinion arrangement S_1. Hence the distance between the slit and collimating lens can be varied. The slit is backed by a light source, *S*. The collimator is rigidly fixed on a stand. When the slit comes exactly into the focal plane of the lens, the beam of light coming out of the lens will be a parallel beam. This is known as the collimating action.

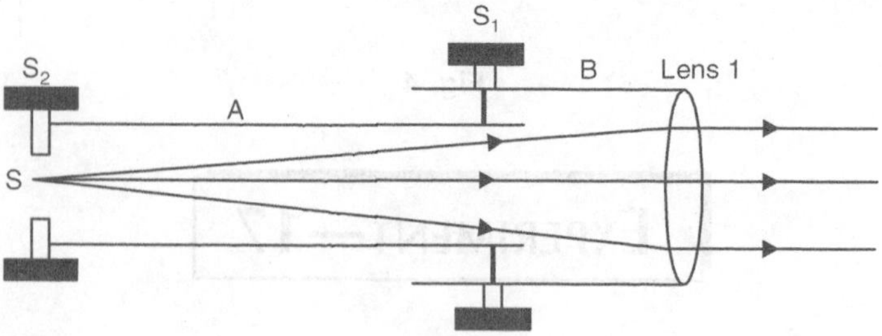

Fig. 17.2

2. **A turntable or prism table.** It consists of a flat circular disc, which can be rotated about a vertical axis, can be raised or lowered and clamped by a side screw. It is fitted with verniers, a concentric circular scale and three levelling screws, which forms the

vertices of an equilateral triangle. The upper surface of the platform has concentric circles and parallel lines for keeping the grating or prism in suitable positions and the angle through which it is rotated can be noted on the circular scale with the help of two verniers fitted 180° apart.

3. The third part is **a telescope** (Fig. 17.3) with cross-wire and a suitable eyepiece to receive the parallel transmitted or reflected beam from the prism or grating. It can also be rotated about the same axis as turntable and it has two verniers attached to it. Tangential screw works only when the clamping screw is tightened. The clamping screw when tightened locks the telescope in position while the tangential screw helps to move the telescope very slowly. The telescope is rigidly fixed to a stand. The whole assembly is mounted on a heavy metallic base having three levelling screws.

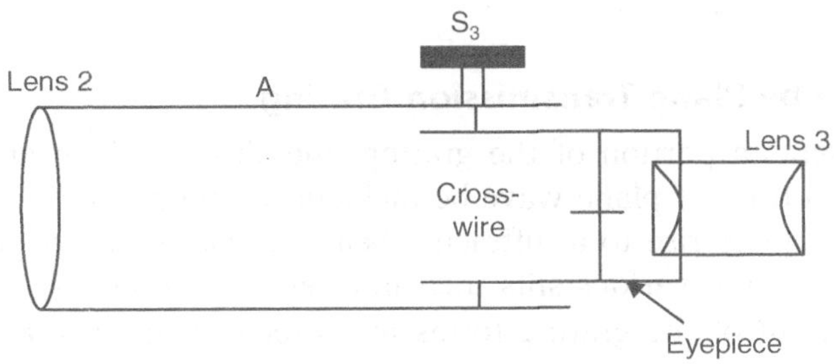

Fig. 17.3

PRINCIPLE

When a beam of white light (Mercury lamp) is passed through a grating it splits up into the beams of constituent colors. The different colored beams when focused on a screen by a converging lens, form an array of colors on the screen because there is a different value of θ for each different color or wavelength for any order of primary maxima (however, the central maxima would still be white). This is called the spectrum of light. Each order will consist of a spectrum so that there will be a number of spectra on either side of the central maxima. The intensity of the higher order maxima will go on decreasing (Fig. 17.4).

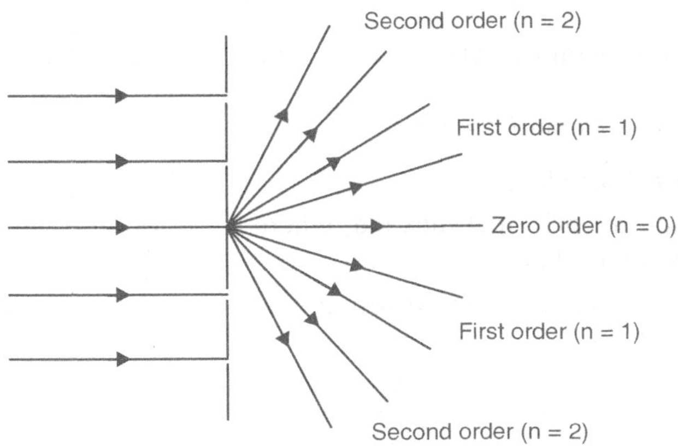

Fig. 17.4

> **Note.** *Grating spectrum differs in many respects from the prism spectrum. In grating spectrum, the amount of spectrum of various colors is proportional to their respective wavelengths hence; spectrum formed by one diffraction grating will be exactly similar to that formed by another except in scale which will obviously depend on the grating element. But this will not be so in the prismatic spectrum for which the dispersion in different parts of the spectrum varies with the characteristics of the material of which the prism is constructed. Since, the refractive index of the optical glass varies more rapidly at violet rather than in red end of the spectrum, hence the prismatic spectrum is much more spread out in proportion at the violet end than at the red. Therefore such a spectrum is called irrational. Another point worth noting is that, in case of prism, the deviation of red color is minimum and the deviation of violet color is maximum. The reverse of it is true for diffraction grating, in which, deviation increases with increase in wavelength.*

THEORY

Formation of Spectra by Plane Transmission Grating

Fig. 17.5 below represents a portion of the grating, the slits of which are perpendicular to the plane of the paper. Let a train of plane wave be incident normally on the grating such as 11′, 12′ and 33′ ... *BB*′. Each slit gives rise to a diffracted beam whose nature depends on the slit width. The diffracted beams from the various slits then interfere with one another to produce the final pattern. A lens at the right of the grating forms in its focal plane a diffracted plane.

Intensity in the case of diffraction grating is given by

$$I = A_0^2 \cdot \frac{\sin^2\alpha}{\alpha^2} \cdot \frac{\sin^2 N\phi}{\sin^2\phi}$$

The expression for intensity contains both diffraction and interference effects. Final diffraction pattern is the resultant of the two as shown in Fig. 17.4 above this experiment. The first factor $A_0^2 \cdot \dfrac{\sin^2\alpha}{\alpha^2}$ may be taken as constant for a slit of definite dimensions. Intensity in final pattern, for fixed *N*, depends upon phase

$$\phi = \frac{\pi}{\lambda}(a + b)\sin\theta$$

Principal maxima are obtained when $\sin\phi = 0$ *i.e.*, $\phi = \pm n\pi$ where $n = 0, 1, 2, \ldots\ldots\ldots$

$$\frac{\pi}{\lambda}(a + b)\sin\theta = n\pi$$

or

$$(a + b)\sin\theta = n\lambda$$

This is the condition for principal maxima, where $(a + b)\sin\theta$ is path difference between the rays starting from various slits.

The position of the spectrum lines corresponding to the wavelength λ in the spectrum of order *n* is given by

$$\sin\theta = \frac{n\lambda}{(a+b)}$$

Since *n* may also be negative integer the spectrum lines also occur at an equal angle θ on other side normal to the grating.

If the incident line contains a number of wavelengths $\lambda_1, \lambda_2, \lambda_3, \ldots$ then in a given order n the spectrum lines corresponding to these wavelengths will be formed at an angle given by

$$\sin \theta_1 = \frac{n\lambda_1}{(a+b)},$$

$$\sin \theta_2 = \frac{n\lambda_2}{(a+b)}, \ldots$$

If $\lambda_1 > \lambda_2 > \lambda_3 \ldots$, Then $\theta_1 > \theta_2 > \theta_3 \ldots$

Thus, the spectrum lines corresponding to larger wavelength occur at larger angle of diffraction on either side of the normal. As the lines become narrow for each order other than zero order ($n = 0$), they will in general be separate from each other. Thus we get the line spectrum of the source on either side of the normal. In the zero order ($n = 0$), all the lines coincide at $\theta = 0$ and we get the central bright line in the direction of the incident ray.

If white light is used the central line will be white. In any other order the spectrum is continuous bright colored patch with violet color of the patch nearer to the central maxima.

Spectrum of the highest order that can be observed with the grating is given by

$$(a + b) \sin \theta = n\lambda \text{ (For normal incidence)}$$

Therefore,
$$n = \left(\frac{a+b}{\lambda}\right) \sin \theta \qquad \ldots(1)$$

The order n will maximum value when $\sin \theta$ is maximum. The maximum value of $\sin \theta$ is 1 when $\theta = 90°$. Therefore, the maximum value of n is given by

$$n_{max} = \frac{a+b}{\lambda} \qquad \ldots(2)$$

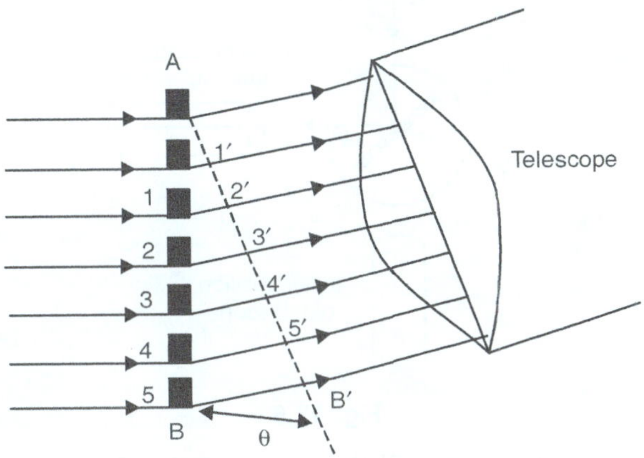

Fig. 17.5

PROCEDURE

1. **Adjustment of the spectrometer :**

Before using the apparatus following adjustments are necessary :

(*i*) First the eyepiece is adjusted so that the cross-wires are in sharp focus (*ii*) The telescope is then adjusted to receive parallel light.

(*iii*) Next the collimator is adjusted so that it gives parallel beam of light.

(*iv*) Finally the optical axes of the telescope and collimator are adjusted so as to lie in the same horizontal line and perpendicular to the axis of rotation of the turn table.

The simplest way to make adjustment (*ii*) is to direct the telescope to a very distant object and to adjust the eyepiece till there is no parallax between the cross-wires and the image of the distant object. The telescope is now adjusted to receive parallel rays (this adjustment is not to be disturbed throughout the experiment). For adjustment (*iii*) the telescope is now directed towards the collimator. The position of the illuminated slit is now adjusted till its image is seen through the telescope as clearly as possible. This ensures that the slit lies in the focal plane of the collimating lens. (This adjustment can also be made by ***Schuster's method*** whose details are given in appendix.)

2. The least count of the spectrometer is determined. The number of lines per inch (L. P. I.), which is written on the grating, is noted and the grating element is evaluated from the relation (as mentioned in observations).
3. The grating is mounted on the prism–table such that it passes through the centre of the table and is made approximately normal to the axis of the collimator by rotating the prism table.
4. The direct image of the slit is then made to coincide with the vertical cross-wire of the telescope. This reading of the telescope is recorded.
5. The telescope is then rotated through 90° from this position and clamped.

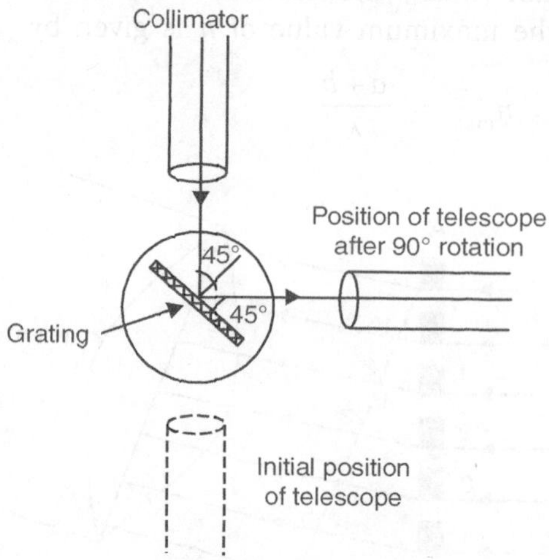

Fig. 17.6

6. The prism table is now rotated till the image of the slit, formed by reflection from the grating is thrown on the cross-wire. In this position the grating is inclined at an angle of 45° or 135° to the collimator axis (as shown in Fig. 17.6).
7. The prism-table is then rotated through 45° from this position. The plane of the grating thus becomes normal to the collimator axis.
8. The telescope is then unlocked and rotated to bring it in line with the collimator to receive the image of the slit on the cross-wire. The reading of the position of the telescope is noted.

9. The telescope is rotated to the right till the image of the slit corresponding to the first order is in the field of view. Now slowly rotating the telescope the first color of the spectrum is made to coincide with the cross-wire. In this position the reading of the scale is noted. The difference between the two positions gives the angle of deviation belonging to the first order spectrum of the specific color. The process is done for all other visible colors.
10. The telescope is rotated further right to get the spectrum belonging to the second order and the step (9) is repeated.

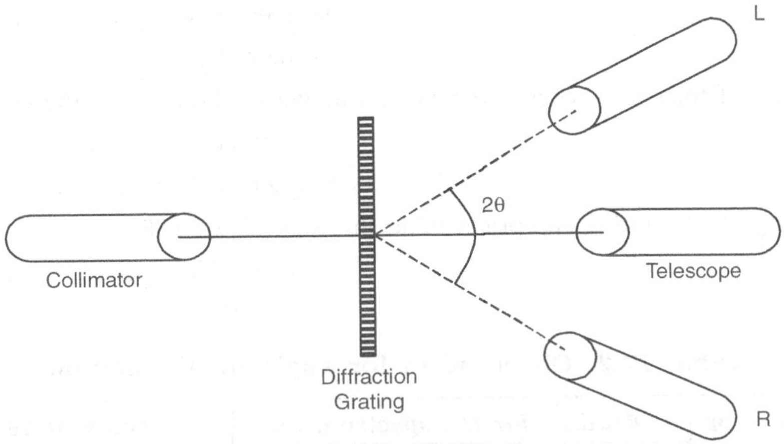

Fig. 17.7

11. Now the telescope is brought to the left of the central line and moved slowly to the left for getting the spectrum belonging to the first order and then second order and step (9), (10) are repeated.
12. The difference in the readings of left and right for the same color gives the value of 2θ for that specific color in that specific order of spectrum.

OBSERVATIONS

Number of lines per inch on the grating, N = ...

Grating element $(a + b) = \frac{2.54}{N}$ cm

= ....cm

For Least count of the spectrometer = ...

Table 17.1

Value of 1 M.S.D.	*Total no. of vernier scale divisions*	*L.C.= 1 M.S.D./total V.S.D.*

Observation for adjustment of the grating :

Least count of the spectrometer $= \dfrac{1\text{ M.S.D.}}{\text{Total no. of V.S.D.}} =$ .... Minutes or degree

Total reading of telescope for direct image vernier $(V_1) =$ ....

vernier $(V_2) =$ ....

Total reading of telescope after rotating it through 90°

vernier $(V_1) =$ ....

vernier $(V_2) =$ ....

Total reading of telescope when the reflected image is obtained on the cross-wire

vernier $(V_1) =$ ....

vernier $(V_2) =$ ....

Total reading after rotating the prism table through 45° or 135°

vernier $(V_1) =$ ....

vernier $(V_2) =$ ....

Table 17.2. Observation for angle of diffraction.

Order of spectrum n	*Color of spectrum*	*Reading for the spectrum on*												*Value of 2θ*			*Value of θ*
		L.H.S.						*R.H.S.*						*Difference in verniers*			
		V_1 (a)			*V_2 (b)*			*V_1 (c)*			*V_2 (d)*			*1st (a ~ c)*	*2nd (b ~ d)*	*Mean*	
		M.S.R.	**V.S.R.**	**TOTAL**	**M.S.R.**	**V.S.R.**	**TOTAL**	**M.S.R.**	**V.S.R.**	**TOTAL**	**M.S.R.**	**V.S.R.**	**TOTAL**				
	Violet																
	Indigo																
	Blue																
	Green																
	Yellow																
	Orange																
	Red																

FORMULA

The wavelength λ can be calculated by the formula

$$(a + b) \sin \theta = n\lambda$$

where 1. $(a + b)$ is the grating element,

2. N is the number of lines drawn per inch on the grating,

3. θ is the angle of diffraction corresponding to the wavelength,
4. λ is the wavelength of the color of light used,
5. n is the order of the spectrum.

CALCULATIONS

Knowing the various quantities, the wavelength of spectral lines of mercury is calculated from the formula :

$$\lambda = \frac{(a + b)\sin\theta}{n}.$$

RESULT

The wavelengths of various spectral lines of mercury are as follows :

Color of spectral lines	*Wavelength 10^{-8} cm or Å*

PRECAUTIONS

1. Do not rub the grating by way of clearing.
2. The slit should be as narrow as possible.
3. The telescope should be focused on the brightest image of the slit while seeing the reflected image.
4. All the preliminary adjustments of the spectrometer must be made before starting the experiment.
5. While taking observations the turn table must remain clamped.
6. Both verniers should be read. Caution must be taken in subtraction. Suppose you get reading of vernier 294° 30′ and 54° 30′ the difference wrongly taken is 240°. It is to be taken correctly as (360° – 294° 30′) + 54° 30′ = 120°.
7. Grating should be mounted at the center of the prism table with its face perfectly normal to the axis of the collimator and the ruled surface facing the telescope.
8. The prism-table must be levelled optically.
9. While recording observations the telescope should be rotated in the same direction in order to avoid backlash error.

CONCLUSION

VIVAVOCE

1. What do you mean by diffraction of light ?
2. What is diffraction grating ?
3. How diffraction grating is constructed ?
4. What do you mean by grating element ? How can you determine its value ?
5. Which light source do you use for your experiment ?
6. How is the spectrum obtained in your experiment when white light source is used ?
7. How many orders of spectrum do you see in your experiment ?
8. Explain Schuster's method adjustment. Why it is done in experiment before taking the observations ?
9. Explain the diffraction pattern obtained in your experiment.
10. What is Fresnel's class of diffraction ?
11. What is Fraunhoffer class of diffraction ?
12. What do you know about the principal maxima of the spectrum ?
13. What do the interference and diffraction of light tell us about the nature of light ?
14. What do you mean by wavefront ?
15. How many types of wavefronts do you know ?
16. If the number of lines drawn on a grating are increased then what happens ?
17. Instead of mercury source if we use sodium vapor lamp, what changes we can observe ?
18. Why do you take readings on two verniers ? Will the two vernier give the same readings ?
19. What is the difference between prism and grating spectra ?

❑❑❑

EXPERIMENT—18

AIM

To determine wavelength of the given Laser Source using diffraction grating.

APPARATUS REQUIRED

He-Ne Laser Source, Diffraction Grating, Grating Stand, White Screen, Ruler.

DESCRIPTION OF THE APPARATUS

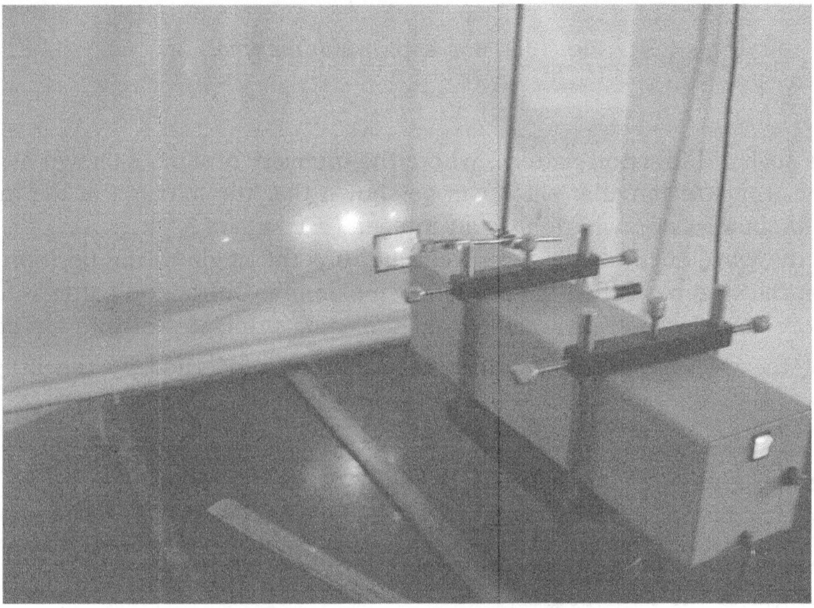

Fig. 18.1. *Laser source with diffraction grating.*

Laser :	Described in sources of light (Experiment–14)
Diffraction Grating :	Described in diffraction (Experiment–17)

PRINCIPLE

Diffraction can be observed when light travels through a hole (in the lab it is usually a vertical slit) whose width *a*, is small. Light from different points across the width of the slit will take paths of different length to arrive at a viewing screen. When the light interferes constructively; intensity maxim appear on the screen.

Increasing the number of slits (while keeping the spacing of adjacent slits constant), gives interference patterns in which the maxima are in the same positions, but progressively sharper and narrower. Because these maxima are so sharp, their angular position, and hence the wavelength of source can be measured to very high precision.

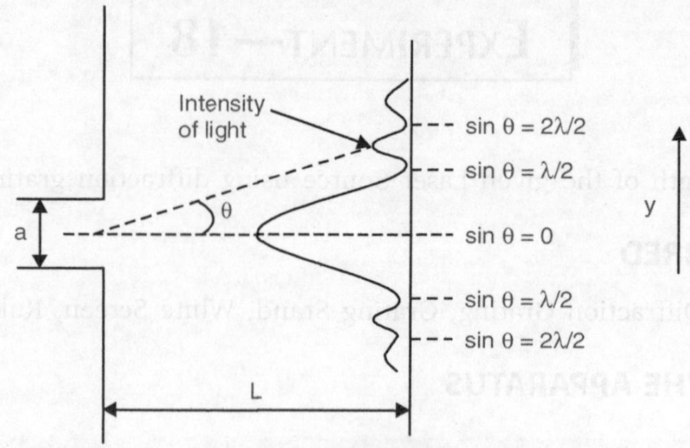

Fig. 18.2. θ *is shown with maxima.*

THEORY

Fig. 18.2 shows such a diffraction pattern, where the intensity of light is shown as a graph placed along the screen. For a rectangular slit it can be shown that the maxima in the intensity pattern fit the formula $a \sin \theta = m\lambda$, where m is an integer ($\pm 1, \pm 2, \pm 3, \ldots\ldots\ldots\ldots\ldots$), a is the width of the slit, λ is the wavelength of the laser light and θ is the angle to the position on the screen. The mth spot on the screen is called the mth order maximum. Diffraction patterns for other shapes of holes are more complex but also result from the same principles of interference.

In case of the grating spectra of different orders are governed by the Bragg's diffraction formula.

$$m\lambda = (a + b) \sin \theta$$

where $(a+b)$ is the grating element, it is given by

$$(a + b) = 1/N$$

where N is the number of lines drawn on the grating m is the order of diffraction maxima and λ is the wavelength of the laser source.

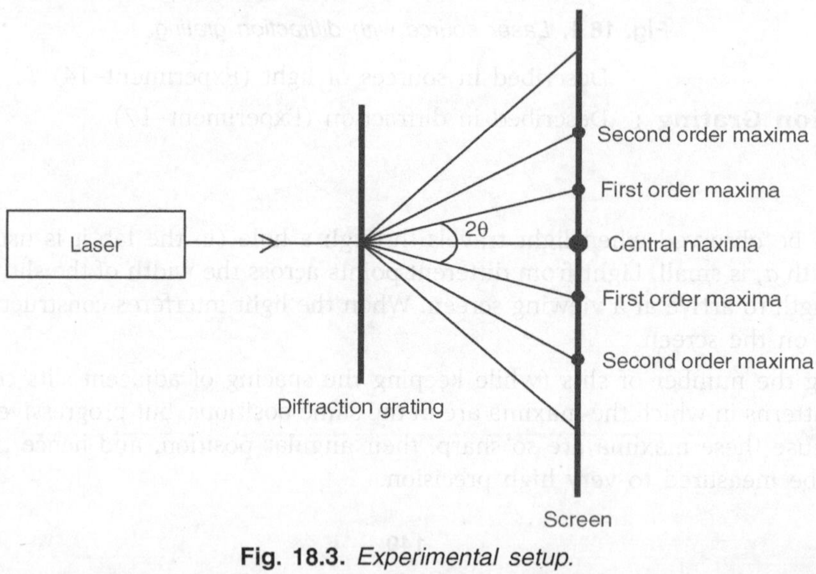

Fig. 18.3. *Experimental setup.*

PROCEDURE

1. Place Laser Source on table.
2. Mount a grating on stand next to laser.
3. Now mount a white screen next to grating (as shown in Fig. 18.3).
4. Adjust the laser such that we can get different diffraction pattern (maxima) on screen.
5. Now central and other maxima are identified.
6. Now measure separation (x) between central maxima and first maxima on either side of the central maxima.
7. Repeat above method for second order maxima and note down all readings in table.
8. Now measure distance (l) between grating and screen.
9. By knowing N (Number of lines drawn on grating) determine value of $(a + b)$ (Grating element which is $1/N$).
10. Repeat all above steps for various distances l.

Note. *This experiment should be done in dark room for better observation and result.*

OBSERVATION

Table 18.1

S.No.	Distance between grating and screen l (cm)	Order of maxima (m)	Distance between central maxima and other maxima x (cm)		Mean x (cm)	Diffraction angle $\theta = \frac{x \cdot 180}{l \cdot \pi}$	sin θ
			LHS	RHS			

CALCULATION

Wavelength of the given Laser Source can be calculated by the formula :

$$m\lambda = (a + b) \sin \theta$$

where

- $(a + b)$ is the grating element and is equal to $1/N$.
- N is the number of lines drawn per inch on the grating.
- θ is the diffraction angle corresponding to the wavelength.
- λ is the wavelength of the Laser Source used.
- m is the order of the maxima.

Use this formula to calculate wavelength of the given laser source for every diffraction order m individually, and then calculate mean wavelength.

RESULT

Wavelength of the given Laser Source is .... nm.

PRECAUTIONS

1. Laser and grating should be properly placed and levelled for getting diffraction patterns.
2. Grating should be handled very carefully.
3. Grating surface is never touched while handling it.
4. Grating should be mounted with its lines perpendicular to the table.
5. Laser beam should not be looked at directly because its beam is highly intense and it may damage your eyes.
6. The laser source is to be switched off when not in use.
7. All the measurements should be taken carefully.

CONCLUSION

VIVAVOCE

1. What is grating ? Why do you get spectral lines when white light passes through it ?
2. How does θ vary with λ ?
3. Will the angular separation of orders be greater for grating with many lines per inch ?
4. Will the angular separation between red and blue rays are greater in the first order or in the second order spectrum ?
5. How many orders of spectrum are possible with a grating of 15000 lines per inch ?
6. What do you mean by diffraction of light ?
7. What is the difference between interference and diffraction ?
8. What do you mean by wavefront ?
9. How laser light is different from ordinary sources of light ?
10. What is grating element ?
11. Explain the working of He-Ne laser.
12. What is the condition for getting maximum intensity on the screen ?

❑❑❑

RESOLVING POWER OF OPTICAL INSTRUMENTS

The ability of an instrument to show two very close objects as separate is called its resolving power. For example, a telescope which is used to see the image of a distant object (say, moon) will be able to separate many craters and mountains present on the moon, if it has good resolving power. But a telescope of poor resolving power will produce only larger image of the moon what we see with our naked eyes.

The image of a point source formed by the objective of a telescope or microscope is not a point image but is a diffraction pattern. If the diffraction patterns of two very close objects are sufficiently small or sufficiently separated from each other the objects can be distinguished from each other. Then they are said to be resolved by the instrument. Resolution of two neighboring point objects is called geometrical resolution and resolution of two close spectral lines is called spectral resolution.

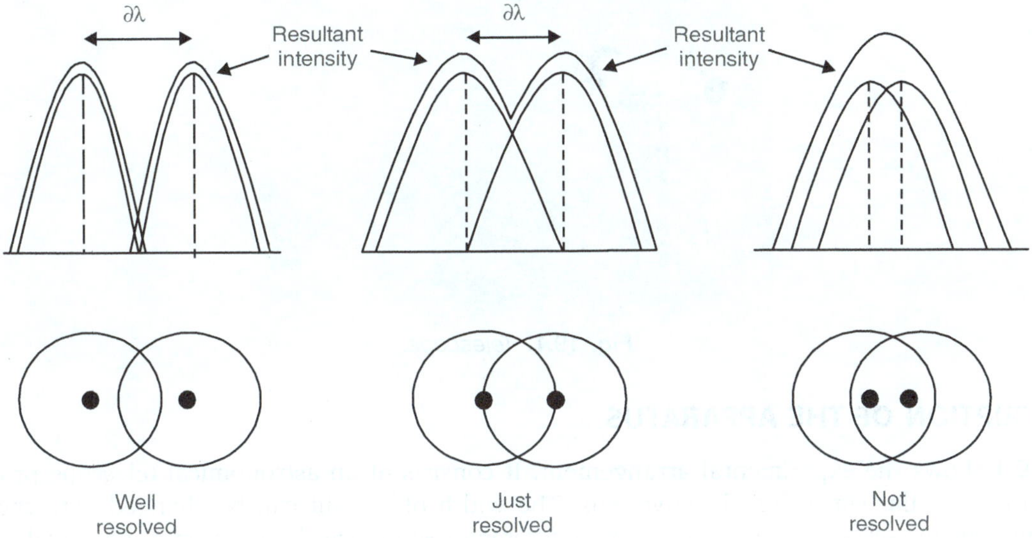

Fig. 1

Now consider a diffraction grating whose purpose is to produce a spectrum. So, resolving power of grating is meant the ability to separate two lines whose wavelengths are very close to each other. The less the difference in wavelengths between two spectral lines which can just be distinguished as separate, greater is the resolving power of the grating. Hence the resolving power of a grating is defined as the ratio of the wavelength of a spectral line to the least difference in wavelength of the next line, which can just be seen as separate.

For objects of size close to that of the wavelength of light, diffraction causes the image to appear as a series of maxima and minima. The images of two closely spaced objects therefore consist of overlapping diffraction patterns. According to Rayleigh criterion. Two spectral lines of equal intensity are just resolved when the principal maxima of the diffraction patterns due to one falls on the first minimum of the diffraction pattern of the other as shown in Fig. 1.

EXPERIMENT—19

AIM

To determine the resolving power of a telescope.

APPARATUS REQUIRED

A telescope, variable slit, a cardboard coated black with equi-spaced and of equal width white strips drawn on it (or a blackened glass plate with equi-spaced and of equal width slits on it and a monochromatic light source such as sodium lamp), vernier callipers, travelling microscope and meter scale.

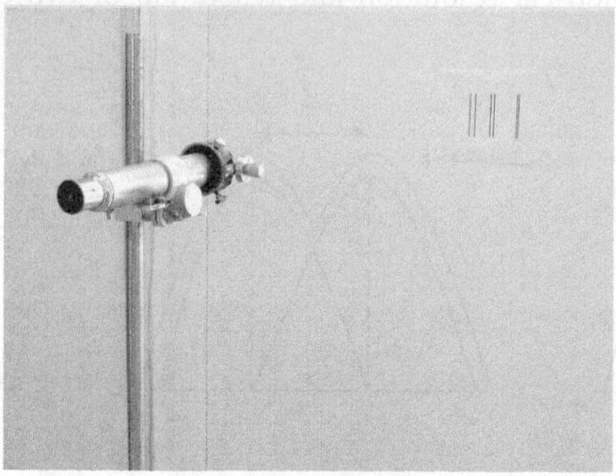

Fig. 19.1. *Telescope.*

DESCRIPTION OF THE APPARATUS

Fig. 19.1 shows the experimental arrangement. It consists of an astronomical telescope provided with a rectangular slit on its objective lens. The width of the slit can be changed with the help of a micrometer screw. The reading of the micrometer screw can be read on a main scale and a circular scale. The telescope is mounted on a vertical stand and is kept horizontally opposite to the cardboard (or the glass plate). Equal spaced and of equal width parallel white strips are drawn on the cardboard (or on the glass plate after blackening it). If glass plate is used, keeping the sodium lamp at its back illuminates it.

PRINCIPLE

The resolving power of an instrument represents the ability to separate the fine details in the object to the eye. It is measured by the angle subtended at the objective by two object points that are just resolved. The Rayleigh criterion states that two objects can be resolved if the first maximum of one image lies on the first minimum of the second.

THEORY

Suppose the telescope is used to view two far objects lying very close to each other. The objective lens LL′ produces not point images but Fraunhoffer diffraction pattern in the focal plane as shown in Fig. 19.2. Obviously, these patterns will lie very close to each other with a large amount of overlapping.

Let the two far objects make an angle θ at the objective lens (shown in Fig. 19.2) the parallel rays *A* comes from one far object and *B* comes from other far object. The first object produces its central maxima at *A′* surrounded by usual diffraction rings.

Similarly, the central maxima of second object are situated at *B′* surrounded by its own set of alternate dark and bright rings. From the diagram, it is clear that angular separation between principal maxima *A′* and *B′* = θ. Now according to Rayleigh, if *A′* is to be distinct from *B′* then first minima of *A′* must fall on *B′* (or first minima of *B′* must fall on *A′*) or it means the path difference *LC* must equal λ, where *L′C* is normal to *LC*.

Now, $LC = LL'.\ \sin\theta = d.\sin\theta$

$= d.\theta$ (If angle θ is small)

where *d* is the diameter of the objective lens.

The condition for resolving becomes

$$d.\theta = \lambda \text{ or } \theta = \frac{\lambda}{d}$$

This gives the limiting angle of resolving in the case of a telescope and is a measure of its resolving power (R.P.). But from the theory of diffraction at a circular aperture, the angular spread of principal maxima $= \dfrac{1.22\lambda}{d}$

∴ The angular separation between the two maxima = Angular spread of principal maxima

or $$\theta = \frac{1.22\lambda}{d}$$

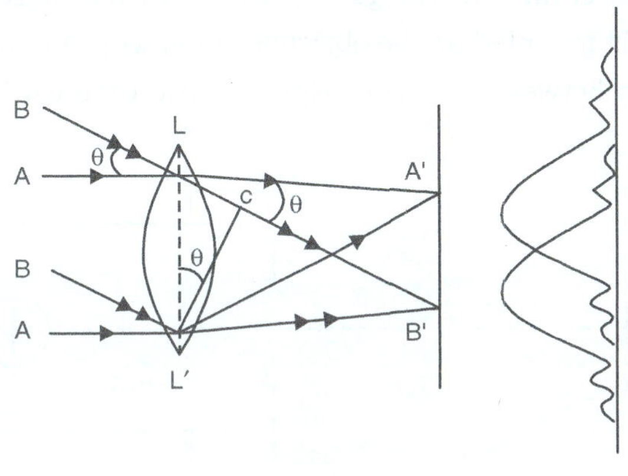

Fig. 19.2

But, by definition, the R.P. is given by the reciprocal of the angular limit of resolution. Hence

$$\text{R.P. of a telescope} = \frac{1}{\theta} = \frac{d}{1.22\lambda} \qquad ...(1)$$

Clearly, smaller the angular separation, greater is the R.P. or in another way, the larger the resolving power, smaller the angle between the two point sources, which can be resolved and hence better the telescope.

Now if the width of slit mounted on the objective lens is '*a*', when it just resolves the two strips at separation '*b*' on the cardboard (or on the glass plate) kept at a distance '*D*', then

The angle subtended by the strips at the slit $= \dfrac{b}{D}$.

= Angular separation between the two principal maxima or the angular spread of principal maxima $= \dfrac{\lambda}{a}$

Hence just at the limit of resolution,

$$\frac{\lambda}{a} = \frac{b}{D}$$

or

$$\lambda = \frac{ab}{D} \qquad \text{...(2)}$$

From equations (1) and (2), resolving power of telescope

$$\text{R.P.} = \frac{dD}{1.22ab} \qquad \text{...(3)}$$

Knowing all the quantities in the above expression, the resolving power of telescope can be calculated.

FORMULA USED

Resolving power of telescope R.P. $= \dfrac{dD}{1.22ab}$ rad^{-1}

where d = Diameter of objective of telescope,

D = Distance of cardboard (or glass plate) from the objective of telescope,

a = Width of slit mounted on the objective of telescope in the position of just resolution,

b = Separation between the two strips drawn on the cardboard (or the glass plate).

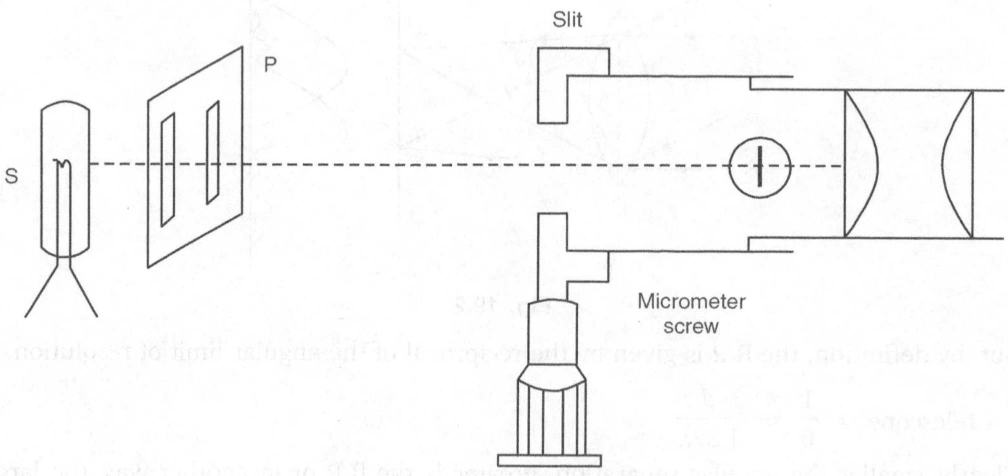

Fig. 19.3. *Experimental arrangement.*

PROCEDURE

1. The cardboard (or glass plate *P*) is clamped on a vertical stand such that the strips drawn on it are exactly vertical as shown in Fig. 19.3.
2. The telescope is clamped horizontally on another vertical stand and is placed at a distance 4–5 meter from the cardboard (or glass plate). Take care that the axis of telescope is horizontal and the height of telescope is in line with the strips drawn on the cardboard.
3. Mount the rectangular variable slit on the objective of telescope such that the slit is vertical and parallel to the strips of the cardboard.
4. Note the least count of the micrometer screw provided with the rectangular slit.
5. Then open the rectangular slit completely and focus the telescope on the strips drawn on the cardboard so that distinct image of strips is seen in the field of view.
6. Now gradually decrease the width of the slit by the micrometer screw till the separate visibility of the two strips just disappears (*i.e.*, the two strips just appear as a single strip). Note the reading of micrometer screw in this position.
7. The micrometer screw is rotated in the same direction till there is complete darkness in the field of view. The reading of micrometer screw gives the width 'a_1' of the rectangular slit in the position of just resolution of the two strips.
8. Now the slit is completely closed and then opened gradually. As we get light in the field of view, the reading of micrometer screw is noted. Then the micrometer screw is turned in the same direction till the two strips just appear to be separated from each other (*i.e.*, the two strips just appear separately). Again the reading of micrometer screw is noted. The difference in the two reading of micrometer screw gives the width 'a_2' of the rectangular slit in the position of just resolution of two strips.
9. The experiment is repeated for the strips of different separation.
10. The least count of travelling microscope is noted and then with its help the distance '*b*' between the two consecutive strips on the cardboard (or on the glass plate) is measured. Note that this distance can also be measured with the help of meter scale if travelling microscope is not given.
11. Note the least count and zero error of the vernier callipers and then with its help find the diameter *d* of the objective lens of the telescope after removing the slit from it.
12. Measure the distance *D* of the cardboard (or the glass plate) from the objective of telescope, using a meter scale (or a measuring tape).
13. Note the wavelength of light used from the standard tables.

OBSERVATION

$$\text{Pitch} = \frac{\text{Distance travelled on the pitch scale}}{\text{Number of rotations of circular head}} = \text{..............cm}$$

(If 2 rotations are given to the circular scale, the distance moved is 1 mm, the Pitch = ½ = 0.5 mm)

Number of divisions on the circular scale = ...

$$\text{Least count of micrometer screw} = \frac{\text{Pitch}}{\text{Total number of divisions on circular scale}}$$

$$= \text{..............................} \text{ cm}$$

Table 19.1. For the width '*a*' of the slit at just resolution.

	Reading of micrometer screw while closing the slit gradually			*Reading of micrometer screw while opening the slit gradually*			*Mean width of slit*
S.No.	*When the resolution just ceases* x_1 *(in cm) M.S.R.+ C.S.R. (C.S.D. × L.C.)*	*When field of view becomes dark* x_2 *(in cm) M.S.R.+ C.S.R. (C.S.D. × L.C.)*	*Width of slit* $a_1 = x_1 \sim x_2$ *(in cm)*	*When light just enters the field of view* x_1' *(in cm) M.S.R.+ C.S.R. (C.S.D. × L.C.)*	*When resolution just starts* x_2' *(in cm) M.S.R.+ C.S.R. (C.S.D. × L.C.)*	*Width of slit* $a_2 = x_1' \sim x_2'$ *(in cm)*	$a = \frac{a_1 + a_2}{2}$ *(in cm)*

Value of one main-scale division of travelling microscope, x = ... cm

Total number of divisions on the vernier scale, n = ...

$$\text{Least count of travelling microscope} = \frac{\text{Value of 1 main scale division}}{\text{Total number of divisions on vernier scale}}$$

$$= \frac{x}{n} \text{ cm}$$

$$= \text{..............} \text{ cm}$$

Table 19.2. For the separation '*b*' between the strips on the cardboard.

S.No.	*Reading of microscope*						*Separation between two strips*
	Position of first strip			*Position of second strip*			
	Main scale reading (M.S.R.) (in cm)	*Vernier scale reading (V.S.R.) (in cm)*	*Total = M.S.R.+ V.S.R.* y_1 *(in cm)*	*Main scale reading (M.S.R.) (in cm)*	*Vernier scale reading (V.S.R.) (in cm)*	*Total = M.S.R.+ V.S.R.* y_2 *(in cm)*	$b = y_1 \sim y_2$
1.							
2.							
3.							

Mean value of b = ...cm

Value of one main-scale division of vernier callipers, x = ... cm

Total number of divisions on the vernier scale, n =.........

$$\text{Least count of vernier callipers} = \frac{\text{Value of one main scale division}}{\text{Total number of divisions on vernier scale}}$$

$$= \frac{x}{n} \text{ cm}$$

= .... cm

Zero error = ± ..... cm

Table 19.3. For the diameter *d* of objective of telescope.

S.No.	*Main scale reading (M.S.R.) (in cm)*	*Vernier scale reading (V.S.R.)*		*Total reading (d) = M.S.R. + V.S.R. (in cm)*
		No. of divisions	*V.S.R. (in cm)*	
1. Along any one direction				
2. Along its perpendicular direction				

Mean observed diameter d = ...cm

True diameter = Observed diameter – zero error (with sign)

= ....cm.

Distance of cardboard from the objective of telescope D = ...cm

Wavelength of light used λ = ... Å = ... cm. (From standard tables)

(Remember that 1 Å = 10^{-8} cm).

CALCULATIONS

Resolving power of telescope

$$\text{R.P.} = \frac{dD}{1.22ab}\ \text{rad}^{-1}$$

Substituting the mean value of ab and the values of d, D in above expression the resolving power can be calculated

$$\text{R.P.} = \frac{dD}{1.22ab} = \text{.........}\ \text{rad}^{-1}$$

RESULT

Experimental value of resolving power of the given telescope = ... rad^{-1}

$$\text{Theoretical value of resolving power} = \frac{1}{\theta}$$

$$= \frac{d}{1.22\lambda} = \text{........}\ \text{rad}^{-1}$$

$$\text{Percentage error} = \frac{\text{Experimental value} \sim \text{Theoretical value}}{\text{Theoretical value}} \times 100\%$$

$$= \text{......}\ \%$$

PRECAUTIONS

1. The axis of telescope must be horizontal.
2. The strips on the cardboard should be vertical.
3. To avoid the backlash error, the micrometer screw must always be turned in one direction.
4. The plane of the slit must be parallel to the cardboard.
5. The width of slit at the position of just resolution must be adjusted carefully.

CONCLUSION

VIVAVOCE

1. What is a telescope ?
2. Explain the construction of a telescope.
3. Why is the aperture of the objective of a telescope large as compared to that of its eyepiece ?
4. Explain what do you mean by magnifying power ?
5. What do you mean by resolving power of an optical instrument ?
6. Is there any relation between magnifying power and resolving power of a telescope ?
7. Can you tell any criterion for resolution which is universally accepted ?
8. What is the difference between a telescope and a microscope ?
9. Can you recognize a telescope or microscope with its outer appearance only ?
10. Which optical phenomenon is observed here ?

❑❑❑

EXPERIMENT—20

AIM

To determine resolving power of a plane transmission diffraction grating.

APPARATUS REQUIRED

Spectrometer, Plane transparent diffraction grating, micrometer slit, mercury lamp etc.

Fig. 20.1. *Spectrometer with grating and micrometer slit.*

DESCRIPTION OF THE APPARATUS

Spectrometer—	described in knowledge of basic measuring instruments in lab.
Plane diffraction grating—	described in experiment–17
Micrometer slit—	described in knowledge of basic measuring instruments in lab.
Mercury lamp—	described in sources of light.

PRINCIPLE

The resolving power of a grating represents its ability to separate spectral lines of two close wavelengths. If the wavelength of two spectral lines are λ and $\lambda + d\lambda$, where $d\lambda$ is minimum difference in wavelength that can just be resolved by grating. Then resolving power of the grating is

$$\text{R.P.} = \frac{\lambda}{d\lambda}.$$

THEORY

Suppose light consisting of two very close wavelengths λ and $\lambda + d\lambda$ is incident normally on a plane transmission grating.

The angular position θ of the principal maximum of wavelength λ in the spectrum of order n is given by the grating equation (as given in experiment–17)

$$(a + b) \sin \theta = n\lambda \qquad ...(1)$$

Differentiating both sides, we get

$$(a + b)\cos \theta .d\theta = n.d\lambda$$

Therefore, dispersive power of the grating in the region of wavelength λ in order n is given by

$$\frac{d\theta}{d\lambda} = \frac{n}{(a+b)\cos \theta} \qquad ...(2)$$

where $d\theta$ is the angle between the two diffracted beams, whose difference in wavelength is $d\lambda$

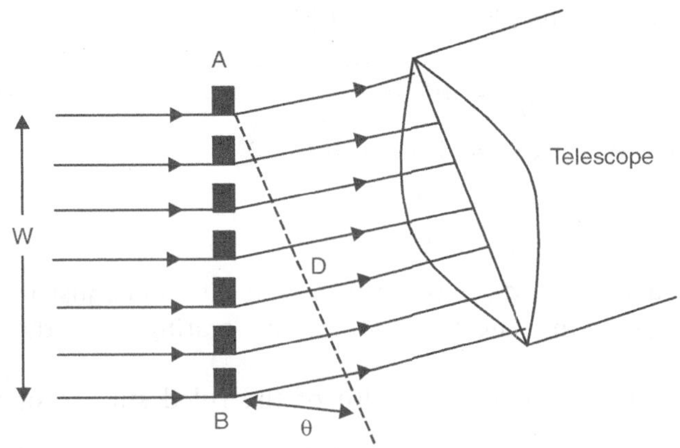

Fig. 20.2

As shown in Fig. 20.2, the light diffracted from the grating enters the objective of a telescope in a grating spectrometer. If the diffracted beam completely fills the objective then width of the beam equals the diameter D of the objective lens. The angular limit of resolution of a telescope objective is given by

$$d\theta = \frac{\lambda}{D} \qquad \text{(As derived in experiment–19)}$$

Now $D = AB \cos \theta = W \cos \theta$

where W is the length of the grating being utilized.

Therefore, $$d\theta = \frac{\lambda}{W \cos \theta}$$

or $$\frac{\lambda}{d\theta} = W \cos \theta \qquad ...(3)$$

From equations (1), (2) and (3), we get

$$\text{R.P.} = \frac{\lambda}{d\lambda} = \frac{\lambda}{d\theta} \times \frac{d\theta}{d\lambda}$$

$$\text{R.P.} = W \cos \theta \times \frac{n}{(a+b)\cos\theta}$$

$$\text{R.P.} = \frac{nW}{(a+b)}$$

$$\text{R.P.} = nN$$

where $N = \dfrac{W}{(a+b)}$ is number of lines on the grating surface being used and n is the order of the spectrum. Hence the resolving power of a grating is given by the simple expression,

$$\text{R.P.} = nN = \frac{(a+b)\sin\theta}{\lambda} \cdot \frac{W}{(a+b)} = \frac{W\sin\theta}{\lambda}$$

FORMULA

The resolving power of grating can be calculated by the formula

$$\frac{\lambda}{d\lambda} = nN = \frac{W\sin\theta}{\lambda} \qquad \text{(For normal incidence)}$$

The maximum value of $\sin\theta = 1$ when $\theta = 90°$

Therefore, $\left(\dfrac{\lambda}{d\lambda}\right)_{max} = \dfrac{W}{\lambda}$

where, λ is the mean wavelength, $d\lambda$ is the wavelength difference just resolved by grating, N is the total number of lines in the effective part of the grating, n is the order of spectrum $= \dfrac{(a+b)\sin\theta}{\lambda}$, $W = N\,(a + b)$ is effective width of the ruled surface or width of the exposed aperture of micrometer slit, $(a + b)$ is grating element, θ is the angle of diffraction

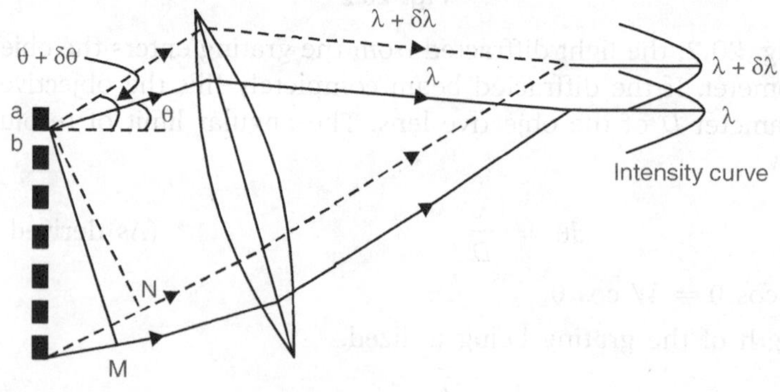

Fig. 20.3

PROCEDURE

1. Set the spectrometer for parallel light rays by Schuster's method (see appendix).
2. Then mount the grating and set it for normal incidence and observe the spectrum on telescope (diffraction pattern).
3. Now align the telescope for first order of spectrum on two yellow D_1 and D_2 lines.
4. Now measure the diffraction angle θ on scale provided in spectrometer for D_1 and D_2 spectral lines with respect to straight light or direct image of slit.

5. Then mount a variable micrometer slit on the collimator lens and keep it wide open.
6. Gradually reduce the aperture in width till the two yellow lines D_1 and D_2 appears as separate. A little more decrease in the width of aperture will make these two lines as one line. Now remove the micrometer slit and measure the width of aperture *i.e., W*.
7. The wavelengths of D_1 and D_2 lines can be observed by spectrum chart of mercury light. The value of $\frac{\lambda}{d\lambda}$ can also be calculated with the help of above values.

OBSERVATION

Least count of the spectrometer :

Value of 1 main scale division = ... degree

Total number of divisions on vernier scale = ...

$$\text{Least count of spectrometer} = \frac{\text{Value of 1 main scale division}}{\text{Total number of divisions on vernier scale}}$$

= ....degree or minutes

Table 20.1. For the angle of diffraction θ.

Order of spectrum	*Color of spectral line*	*Position of telescope on the left side of direct image of slit*			*Position of telescope on the direct image of slit*			*Angle of diffraction* $\theta = a \sim b$ *(in degree)*
		Main scale reading (M.S.R.)	*Vernier scale reading (V.S.R.)*	*Total = M.S.R. +V.S.R. 'a'*	*Main scale reading (M.S.R.)*	*Vernier scale reading (V.S.R.)*	*Total = M.S.R.+ V.S.R. 'b'*	
$n = 1$	Yellow (D_1)							
	Yellow (D_2)							
$n = 2$	Yellow (D_1)							
	Yellow (D_2)							

Mean θ = ...

Least count of the micrometer screw :

$$\text{Pitch} = \frac{\text{Distance travelled on the pitch scale}}{\text{Number of rotations of circular head}} = \text{............cm}$$

(If 2 rotations are given to the circular scale, the distance moved is 1 mm, the Pitch = ½ = 0.5 mm)

Number of divisions on the circular scale = ...

$$\text{Least count of micrometer screw} = \frac{\text{Pitch}}{\text{Total number of divisions on circular scale}} = \text{........ cm}$$

Table 20.2. For reading of micrometer screw for aperture width *W* of slit.

Order	***Main scale reading (M.S.R.) (in cm)***	***Circular scale reading (C.S.R.) = number of division coinciding the base line × least count (in cm)***	***Total reading = M.S.R. + C.S.R. (in cm)***
$n = 1$			
$n = 2$			

CALCULATION

$$\text{The resolving power of grating} = \frac{W \sin \theta}{\lambda}$$

Substituting the values of W, θ and λ resolving power can be calculated.

RESULT

$$\text{Theoretical value of resolving power of grating} = \frac{\lambda}{d\lambda} = \text{............}$$

$$\text{The experimental value of resolving power of grating} = \frac{W \sin \theta}{\lambda} = \text{..............}$$

PRECAUTIONS

1. Slit should be clean at edges and narrow.
2. The proper setting of spectrometer, its telescope and collimator must be done.
3. The clamping should be proper while taking the observations.
4. Both the vernier should be read remembering the marking on its circular main scale division. The main scale is graduated as 0° (360°), 90°, 180°, 270° and 360° or 0°. Caution must be taken in subtraction. Suppose you get reading of verniers 294° 30′ and 54° 30′ the difference wrongly taken is 240°. It is to be taken correctly as (360 – 294° 30′) + 54° 30′ = 120°.
5. Grating must be set for normal incidence (as given in Fig. 17.6, Experiment-17).

CONCLUSION

VIVAVOCE

1. What is Rayleigh's criterion for the resolution of spectral lines ?
2. What is the difference between resolving power and magnifying power of an optical instrument ?
3. What is resolving power of a grating ?
4. Differentiate optical and geometrical resolution.
5. What is the position of objective lens in telescope ?
6. What do you mean by diffraction of light ?
7. Explain the setting of a spectrometer for normal incidence of light.
8. What do you mean by pitch of the micrometer slit ?
9. Why two veniers are provided in the spectrometer ?
10. What is angular width of principal maxima ?

*For more questions please refer to experiment-17.

❑❑❑

EXPERIMENT—21

AIM

To determine the Beam divergence of a laser beam.

APPARATUS REQUIRED

Helium-Neon/Diode Laser, White Screen, Ruler and measuring tape for measuring distances.

Fig. 21.1. *Diverged laser beam.*

DESCRIPTION OF THE APPARATUS

Laser : Described in sources of light before experiment–14.

A white screen is used here to obtain the beam divergence of laser beam. Ruler and measuring tape are used to measure the distance of laser from screen, laser output diameter and diameter of spot on the screen.

PRINCIPLE

Light emitted by normal light sources is non-directional and less intense because they emit light in all directions. Its intensity decreases with distance. But in case of laser it is different. The divergence of a laser beam is primarily due to diffraction effects. The radius of the cross-section of the laser beam, *i.e.,* the spot size is about a few millimeters. As the beam propagates, the spot size increases and the rate of increase in the width is proportional to the wavelength and inversely proportional to the initial width of the beam.

THEORY

One of the main advantages of laser is that it produces unidirectional and intense beam whose edges are parallel. Generally it is assumed that laser emits beam, which have zero divergence. But in practical condition it is an unobtainable task, which is difficult to achieve. Any deviation from parallelism will cause the laser beam to diverge and spread out its energy. Therefore, Laser beam has divergence but it is extremely low.

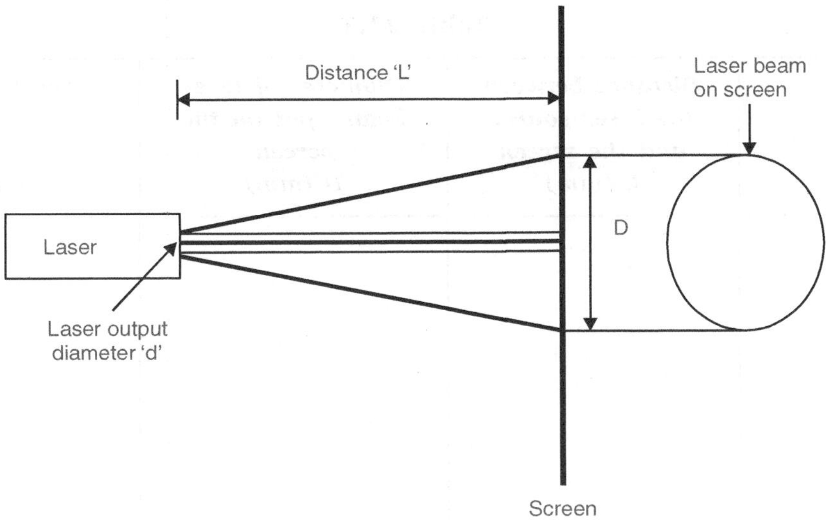

Fig. 21.2

When a screen is placed in front of laser source and beam is allowed to fall on the screen, a circular spot can be seen on the screen (as shown in Fig. 21.2). When the screen is moved far or near from laser source, the size of the circular aperture increases or decreases respectively.

If D is the diameter of circular spot on the screen and L is the distance of the laser source from the screen, then divergence θ_0 is given by the relation

$$\theta_0 = D/L \text{ (radian)}$$

PROCEDURE

1. Put the He-Ne laser on the table.
2. Place the screen, directly in front of the He-Ne laser, so that the light from the laser is perpendicularly incident on the screen.
3. Place the screen at different positions, first start close to the laser source (say 50 cm).
4. Measure the Laser source output (d), from where laser beam is emitted out.
5. Measure and note the distance of the screen from the laser (L).
6. Measure and note the diameter (D) of the spot on the screen.
7. Now move the screen farther and again measure distance and diameter for this position. Note this reading on the table.
8. Repeat the above process for different positions of the screen. Record in table.
9. For more accurate measurement of diameter subtract the laser output (d) from the measured value (D), and call it the new value (D^*).
10. Draw a graph of D^* vs L.
11. Find the slope of the graph.

OBSERVATION TABLE

Diameter of laser source output d = ...mm

Table 21.1

S.No.	Distance between the laser source and the screen L (mm)	Diameter of laser beam spot on the screen D (mm)	Corrected diameter D* (mm) = D – d

CALCULATIONS

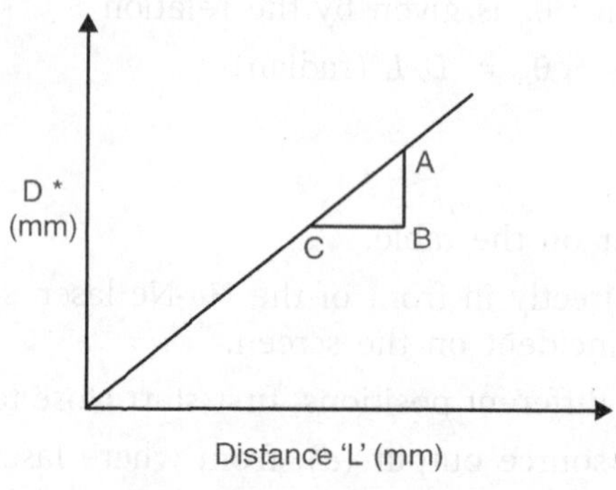

Fig. 21.3

Draw the graph of diameter of the spot on the screen D^* (mm) as a function of the distance between laser source and screen L (mm).

Slope of the graph = AB/BC

= ....

Angle of divergence θ_0 = Slope = ... (radians).

= .... × $(180/\pi)$ degrees.

Note. *1 radian = $(180/\pi)^\circ$ = 57.30°.*

RESULT

The angular divergence of the given Laser beam is .... degrees.

PRECAUTIONS

1. Laser and screen should be properly placed and levelled for getting proper circular spot.
2. Laser beam should not be looked at directly because its beam is highly intense and it may damage your eyes.
3. The laser source is to be switched off when not in use.
4. All the measurements should be taken carefully.

CONCLUSION

VIVAVOCE

1. Explain how He-Ne laser source works ?
2. What do you mean by beam divergence of laser ?
3. Why is laser light coherent ?
4. Why is the divergence of Laser beam extremely low ?
5. Which laser is using in your experiment ? What is the wavelength of the light emitted by it ?
6. Why is the laser beam highly intense ?
7. Explain the working of He-Ne laser.
8. Explain the working of diode laser.
9. How is beam divergence measured in this experiment ?
10. What do you mean by gas laser ?

❑❑❑

EXPERIMENT—22

AIM

To determine the index of refraction 'n' of the given glass slab with the help of laser beam.

APPARATUS REQUIRED

Helium-Neon/Diode Laser, Small rectangular Glass Slab with magnetic mount, Vertical Steel Mount, Graph Paper, Small Ruler and Protactor.

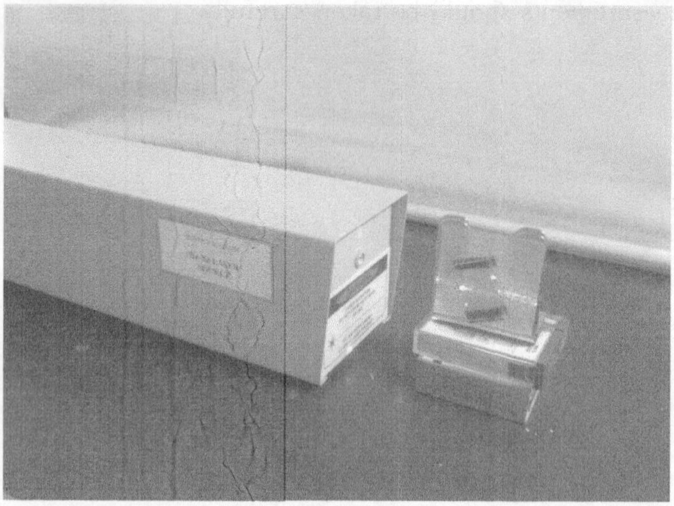

Fig. 22.1. *He-Ne laser source and rectangular glass slab with magnetic mount.*

DESCRIPTION OF THE APPARATUS

In this experiment a He-Ne laser source (described in experiment–14) has been taken. A rectangular glass slab is used as a transparent material, whose refractive index is to be measured. It is attached over a graph paper with magnetic mount.

PRINCIPLE

The refractive index of a substance measures how the substance affects light travelling through it. It is equal to the speed of light in a vacuum divided by the speed of light in that substance. When light travels between two materials with different refractive indexes, it bends at the boundary between them.

If the angle of incident is i and angle of refraction is r then from Snell's law the index of refraction n of the given glass is given by

$$n = \frac{\sin i}{\sin r}$$

THEORY

When light travels from one transparent medium to another medium, a change of direction as well as speed of light is observed. This bending of light is called refraction of light. When light travels from rarer to denser medium, it bends towards normal and when light travels from denser to rarer medium it bends away from normal. This phenomenon can be well-explained with the help of laser beam. If the angle between the laser beam and glass surface is less than 90°, bending of the laser beam occurs at that point. The bending depends on the incident angle and the index of refraction of the glass.

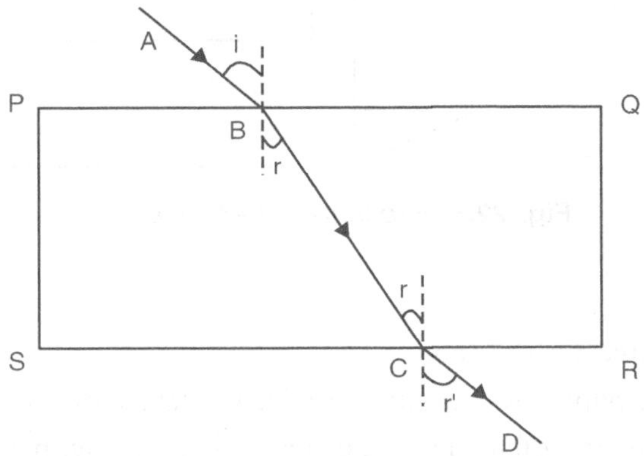

Fig. 22.2. *Refraction through glass slab.*

PQRS is a rectangular slab of glass with uniform thickness. Let a laser beam *AB* travelling through air be incident on its upper surface at *B* (Fig. 22.2). This beam gets refracted and moves along *BC* within the slab. When it reaches the lower surface, it gets refracted again as it re-enters air. The cross-section of the slab and the path of the ray are given in Fig. 22.2, *PQ* and *SR* are the upper and lower surface of the slab. Here $PS = QR$ (thickness of the slab).

For the refraction at *B*, Let *i* and *r* be the angle of incident and angle of refraction. If *n* is the refractive index of the material of the slab,

$$\frac{\sin i}{\sin r} = n \qquad \ldots(1)$$

For the refraction at *C, BC* is the incident ray and *CD* is the refracted ray. Since the surfaces *PQ* and *SR* are parallel, the normal to them are also parallel. Let the angle of refraction be r' at *C*.

Hence from Snell's law $\dfrac{\sin r}{\sin r'} = \dfrac{1}{n}$

or $\sin r' = n \sin r \qquad \ldots(2)$

From (1) and (2) $r' = i$

This means that the beam *CD* is parallel to the beam *AB*. Thus, when a beam of laser light passes through a transparent slab with parallel faces, the ray is displaced parallel to itself and the amount by which the ray is displaced is proportional to the thickness of the slab.

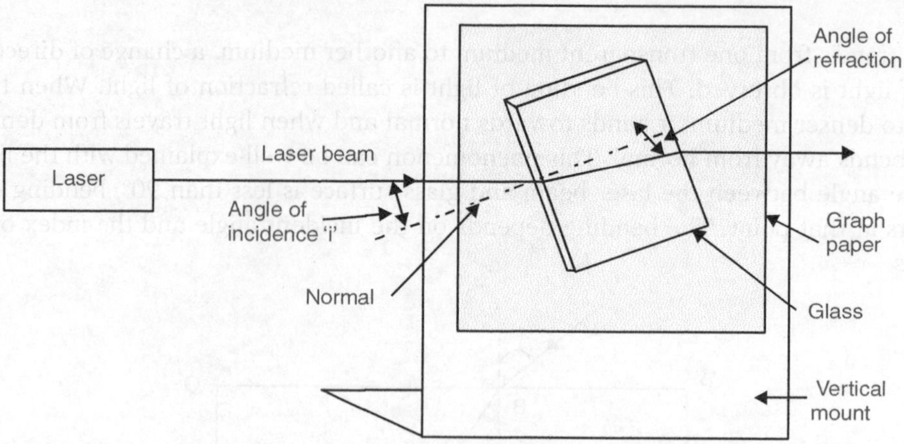

Fig. 22.3. *Experimental arrangement.*

PROCEDURE

1. Place laser source on the table.
2. Attach a graph paper over a surface of a vertical mount.
3. Now attach a rectangular glass slab over the graph with magnetic mount.
4. Place this glass setup in front of the given laser source, so that the laser beam can pass through this rectangular glass (as shown in Fig. 22.3).
5. Rotate the glass so that the laser beam enters the glass at an angle less than 60° (say 30°) and emerges out from the opposite side.
6. On the graph behind the glass, carefully mark points where the laser beam enters and emerges out the glass. Also mark the outline of the glass.
7. Take the graph paper out and measure the angle of incidence (***i***), which is formed by the laser beam and the normal to the surface of the glass.
8. Also measure the angle of refraction (***r***) formed by the normal to the surface and the refracted laser beam inside the glass.
9. Repeat the above procedure for different angle of incidence by rotating the glass. Note all observations on the table.
10. For each angle of incidence, calculate the Index of refraction of the glass using Snell's Law.

OBSERVATION TABLE

Table 22.1

S.No.	*Angle of incidence (degree)*	*Angle of refraction (degree)*	*Index of refraction 'n'*

Mean *n* ....

CALCULATION

The index of refraction 'n' can be calculated by the Snell's formula :

$$n = \frac{\sin i}{\sin r}$$

where i = Angle of incidence.

r = Angle of refraction.

n = Index of refraction.

RESULT

The index of refraction of the given glass slab is found to be ....

PRECAUTIONS

1. Laser source should be properly placed and levelled on the table.
2. Glass should be mounted properly.
3. Angle of incidence should be less than 60° for proper observation.
4. Laser beam should not be looked at directly because its beam is highly intense and it may damage your eyes.
5. The laser source is to be switched off when not in use.
6. All the measurements should be taken carefully.

CONCLUSION

VIVAVOCE

1. What is Snell's law ?
2. Differentiate between reflection and refraction of light.
3. What do you mean by rarer and denser medium ?
4. What are the special characteristics of laser light ?
5. What is coherence of light ? Explain special coherence and temporal coherence.
6. What do you mean by refractive index ?
7. Explain how He-Ne laser source works ?
8. What are the applications of laser source ?
9. What do you mean by transmission medium ?
10. Why are you using laser source here ?

❑❑❑

EXPERIMENT—23

AIM

To determine the Numerical Aperture (NA) of Optical Fiber.

APPARATUS REQUIRED

Experimental Kit Fiber Link-A (make FALCON), Fiber cable, Fiber holding fixture, traveling microscope, graph paper.

DESCRIPTION OF THE APPARATUS

Optical fibers are glass or plastic wave-guides designed to propagate light signal along their length. A single optical fiber can be as thin as human hair. It is normally cylindrical in shape. An optical fiber has in general three coaxial regions namely core, cladding and sheath (Fig. 23.1).

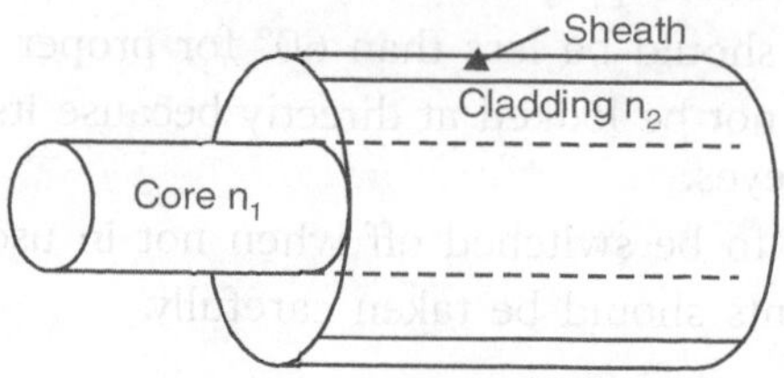

Fig. 23.1. *Optical fiber.*

The innermost part of the fiber that propagates the light signal is called core. The core is surrounded by another coaxial region having refractive index slightly lower than core is called cladding. The purpose of the cladding is to confine light signal within core by successive total internal reflection. Sheath is the outer most region of the fiber. It is generally a plastic coating. It provides safety to fiber from atmosphere and also provides strength to the fiber.

Fig. 23.2. *The optical fiber experiment kit.*

Fig. 23.2 shows the experimental set up for finding numerical aperture of optical fiber. One end of the fiber is connected with a Kit fiber link—A and other end is inserted in fiber holding fixture. When the power supply is on, the signal launched into the fiber in the form of light using LED and emerges out at the other end of the fiber in the form of a cone. This cone is called acceptance cone of the fiber (Fig. 23.3). The output signal strikes on graph paper fixed on the base of fiber holding fixture as red colored spot. A traveling microscope is placed in front of the fiber holding fixture to measure the height of the acceptance cone.

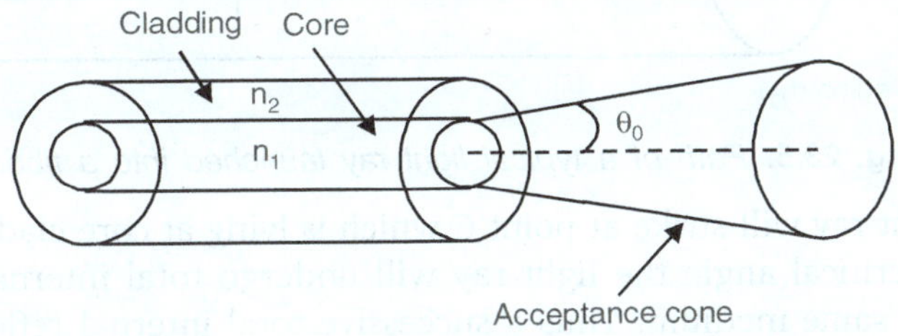

Fig. 23.3. *Acceptance cone of the fiber.*

PRINCIPLE

The optical fibers work on the principle of total internal reflection. When light enters one end of a glass fiber under a right condition, it gets successive total internal reflection and finally emerges out at the other end of the fiber with minimum loss. The main function of the fibers is to accept more signals and transmit it at the other end of the fiber. The light gathering ability of the fiber is called numerical aperture of the fiber.

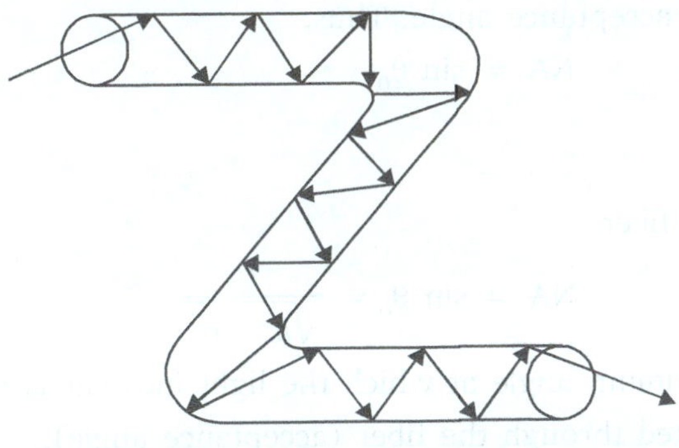

Fig. 23.4. *Reflection from the inside of the core wall.*

THEORY

To understand the experiment let us consider an optical fiber having core of refractive index n_1 surrounded by a cladding of material with a lower refractive index n_2 (Fig. 23.5). Light enters at one end of the fiber from a launch region with a refractive index n_0. If the launch region is air, then $n_0 = 1$.

Let us consider a light ray *AB* is launched at one end of optical fiber with angle of incidence *i* and refracted with angle *r*.

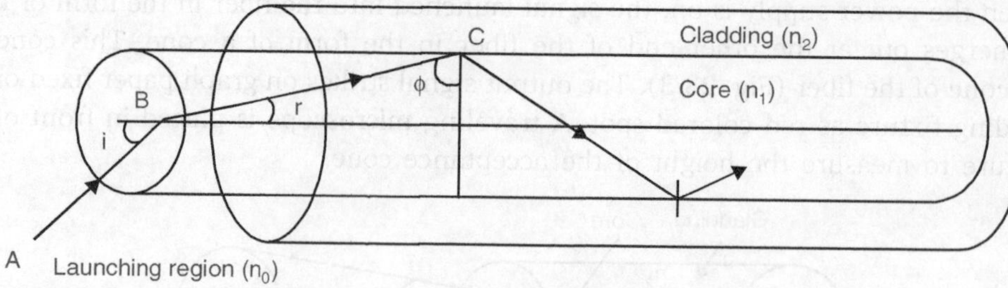

Fig. 23.5. *Path of a typical light ray launched into a fiber.*

This refracted light ray will strike at point C which is lying at core cladding interface at angle ϕ. If ϕ is greater than critical angle the light ray will undergo total internal reflection at point C and reflect back to the same medium. Thus a successive total internal reflection of the light rays takes place and the signal is received at the other end of the fiber.

If we increase angle of incident i at point B, the angle ϕ at point C decreases. And for a maximum value of i, ϕ is equal to critical angle. Beyond this limit ϕ will become less than critical angle and light will escape from the fiber.

The maximum value of angle of incident, for which the light is confined within the fiber, is called acceptance angle of the fiber. It is denoted by θ_0.

The light rays contained within the cone having full angle $2\theta_0$ are accepted and transmitted along the fiber. This cone is called acceptance cone.

Numerical aperture is the light gathering ability of the fiber. The numerical aperture (NA) is defined as the sine of the acceptance angle. Thus,

$$\text{NA} = \sin \theta_0$$

FORMULA USED

Numerical aperture of the fiber

$$\text{NA} = \sin \theta_0 = \frac{r}{\sqrt{h^2 + r^2}}$$

where θ_0 = The maximum angle at which the light incident is properly transmitted through the fiber (acceptance angle).

r = Radius of the illuminated circular patch of light on the screen.

h = Height of the fiber from base.

PROCEDURE

1. Connect one end of the fiber into fiber link A and other end to the fiber holding fixture.
2. Fix the graph paper facing the fiber. Adjust the fiber such that its cut face is perpendicular to the axis of the fiber.
3. Gently tighten the screw and thus fix the fiber in the place.
4. Now measure the distance of the fiber from the screen 'h' with the help of travelling microscope.

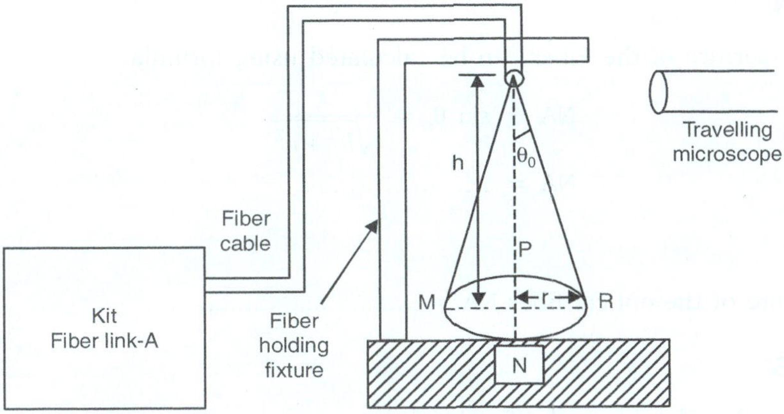

Fig. 23.6. *Experimental arrangement for numerical aperture measurement.*

5. Switch on the power supply.
6. Now observe the illuminated circular patch of light on the screen.
7. Measure the vertical and horizontal diameters *MR* and *PN* indicating in Fig. 23.6.
8. Mean radius is calculated using the formula

$$r = \frac{(MR + PN)}{4}$$

9. Repeat steps 4, 5 and 6 for different values of '*h*'.
10. Find the numerical aperture of the fiber by putting the values of '*r*' and '*h*' in the formula.

OBSERVATIONS

$$\text{Least count of travelling microscope} = \frac{\text{Value of one division on main scale}}{\text{Total number of divisions on vernier scale}}$$

$$= \text{..........................} \text{ cm}$$

Table 23.1. For measuring height of fiber tip from base.

S.No.	Radius of the circular patch 'r' (cm)	Microscope reading of the fiber tip			Microscope reading of the edge of the base (screen)			Height of the fiber tip $h = x_1 - x_2$ (cm)
		M.S.R (cm)	V.S.R. (cm)	TOTAL x_1 (cm)	M.S.R (cm)	V.S.R. (cm)	TOTAL x_2 (cm)	
1.	r_1 =							h_1 =
2.	r_2 =							h_2 =

CALCULATION

The Numerical Aperture of the Fiber can be calculated using formula

$$NA = \sin\theta_0 = \frac{r}{\sqrt{h^2 + r^2}}$$

$$NA = ...$$

RESULT

Numerical aperture of the optical fiber NA = ...

PRECAUTIONS

1. Fiber cable should not be fixed tightly.
2. The diameter of circular patch should be measured exactly.
3. Damaged fiber should not be cut using blade but with proper tool.

CONCLUSION

VIVAVOCE

1. What is optical fiber ?
2. On which phenomena they work ?
3. Explain total internal reflection.
4. How light signal propagates along optical fiber ?
5. What is numerical aperture of an optical fiber ?
6. What is acceptance angle of an optical fiber ?
7. What is acceptance cone ?
8. How many parts are there in optical fiber ?
9. What is the working of core, cladding and sheath of an optical fiber ?
10. What are the advantages of optical fiber over copper wires ?
11. How do you find numerical aperture of the given optical fiber ?
12. What do you mean by critical angle ?

❑❑❑

EXPERIMENT—24

AIM

To determine the specific charge (e/m) of electron by Thomson's Method.

APPARATUS REQUIRED

Cathode ray tube, Power supply with voltmeter, a pair of bar magnet, wooden stand with scale, compass box with stand.

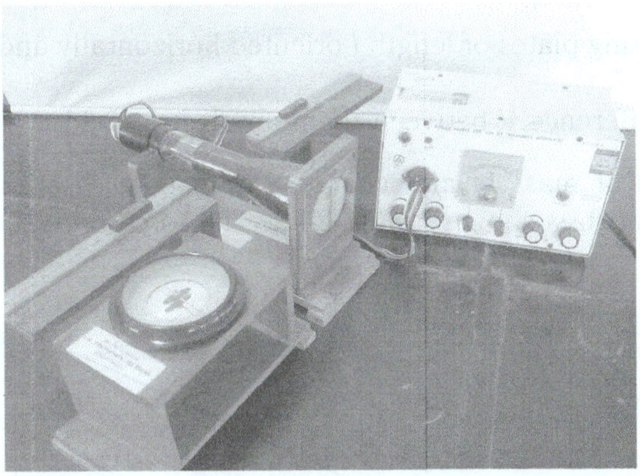

Fig. 24.1. *Setup for e/m of electron by Thomson's method.*

DESCRIPTION OF THE APPARATUS

It consists of a highly evacuated glass tube, fitted with electrodes, as shown in Fig. 24.1. The cathode rays are produced from cathode via thermionic emission (heating a tungsten filament electrically produces electrons). Then they are passed through fine slits in two anode, called the focusing anode and the second anode is called accelerating anode. When a high potential difference is applied between cathode and anode through built in power supply, these electrons are accelerated and collimated by the slits into a narrow parallel beam. By applying a high potential difference between two horizontally parallel deflecting plates, the deflecting electric field can be produced perpendicular to the electronic motion. The magnetic field is applied by permanent magnets (or electromagnets) with its direction at right angles to the cathode rays and the electric field. The electric field, magnetic field and electron path are mutually perpendicular.

When electric and magnetic fields are not switched on, the electrons follow straight path and strike the fluorescent screen coated with phosphor on the opposite side and produce a glow at the centre of the screen. Intensity of the spot is related with number of focused electrons reaching at the centre. When the electric field alone is switched on, the electrons are deflected upward and on the other hand when the magnetic field alone is switched on, the electrons are deflected downward. They are passed through electric and magnetic field simultaneously. Finally they fall on fluorescent screen. The amount of deflection can be noted on the scale attached on the screen. The compass box used to measure the intensity of the applied magnetic field.

PRINCIPLE

It works on the principle of motion of charged particles in uniform electric and magnetic field. If the fields are adjusted such that the electric force balances the magnetic force acting on an electron beam having electron velocities spread around a central value v, then those electrons are not deflected and continue to travel along a straight path. Subsequently they strike at center of the screen. Electron moving with a lesser velocity will get deflected upward and those moving with a higher velocity will get deflected downward. The walls of the vacuum tube absorb the electrons deflected away.

THEORY

Let P and Q are two deflecting plates of length l oriented horizontally and separated by a distance d.

By applying a potential difference V between them an electric field $E = \frac{V}{d}$ is produced in vertical direction. Let an electron of mass m and charge e moving with a horizontal velocity v enters between the plates at point A as shown in Fig. 24.2. The force on electron in Y-direction, due to electric field is $F_E = eE$.

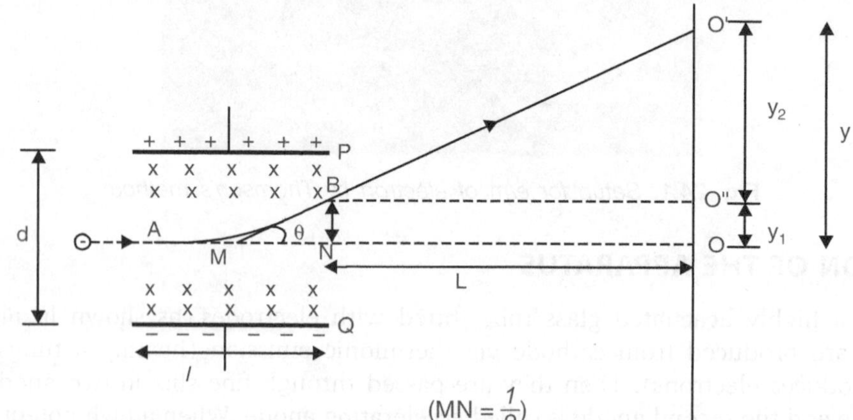

Fig. 24.2

As the force along X-direction is zero, the electron continues to travel with uniform velocity v along the X-direction. So distance travelled by electron in the X-direction is

$$l = vt \qquad ...(1)$$

And the displacement in Y-direction at point B

$$y_1 = \frac{1}{2}at^2$$

or

$$y_1 = \frac{1}{2}\left(\frac{eE}{m}\right)\left(\frac{l}{v}\right)^2 = \frac{1}{2}\frac{eEl^2}{mv^2} \qquad ...(2)$$

The angular displacement of electron in the electric field is given by

$$\tan\theta = \left(\frac{dy_1}{dl}\right) = \frac{eEl}{mv^2} \qquad ...(3)$$

After emerging out of the electric field, the electron travels in a straight line at an angle θ from its initial direction and strikes the screen at point O'. So additional deflection on the screen

$$y_2 = L \tan \theta = \frac{eElL}{mv^2} \quad ...(4)$$

Hence total vertical deflection obtained on the screen

$$OO' = y = y_1 + y_2$$

$$y = \frac{eEl^2}{2mv^2} + \frac{eElL}{mv^2}$$

$$y = \frac{eEl}{mv^2}\left(\frac{l}{2} + L\right)$$

$$\frac{e}{m} = \frac{v^2 y}{El\left(\frac{l}{2} + L\right)} \quad ...(5)$$

For determination of v:

When a magnetic field is applied perpendicular to electronic motion, the electrons experience a Lorentz force $F_L = evB$. If the magnetic field is so adjusted such that the deflection OO' of electron beam produced by electric field is annulled by the magnetic field, *i.e.*, electron beam strikes at point O again. In this condition, the electric force F_E is equal and opposite of the force F_L due to the magnetic field B, *i.e.*,

$$F_E = F_L$$

$$eE = evB$$

$$v = \frac{E}{B} \quad ...(6)$$

Substituting the value of v from equation (6) in equation (5),

$$\frac{e}{m} = \frac{(E/B)^2 y}{El\left(\frac{l}{2} + L\right)} = \frac{Ey}{B^2 l\left(\frac{l}{2} + L\right)} \quad ...(7)$$

For the measurement of magnetic field B, the CRT is withdrawn and the compass box is placed in between the bar magnets. If the deflection of the magnetic needle in the compass box is θ, then from tangent law,

$$B = H \tan \theta$$

Therefore, from equation (7),

$$\frac{e}{m} = \frac{Ey}{(H \tan \theta)^2 l\left(\frac{l}{2} + L\right)} \quad ...(8)$$

Also $E = \frac{V}{d}$, where V is the potential difference applied between the plates, d is the separation between the plates. Therefore, equation (8), becomes

$$\frac{e}{m} = \frac{Vy}{d(H \tan \theta)^2 l\left(\frac{l}{2} + L\right)} \quad ...(9)$$

or $$\frac{e}{m} = K\frac{Vy}{\tan^2\theta} \text{ Coulomb/kg (in SI units)} \quad ...(10)$$

where, $$K = \frac{1}{dH^2 l\left(\frac{l}{2}+L\right)} = \text{a constant}$$

FORMULA USED

The specific charge of electron

$$\frac{e}{m} = K\frac{Vy}{\tan^2\theta} \text{ Coulomb/kg,}$$

where $$K = \frac{1}{dH^2 l\left(\frac{l}{2}+L\right)}.$$

V = Voltage difference applied on the deflection plates in volts.
y = Displacement of the spot on the screen in meter.
l = Length of deflection plate in meter.
L = Distance of the screen from the edges of the deflection plates.
H = Horizontal component of earth's magnetic field.
d = Separation between the deflection plates in meter and
θ = The deflection of magnetic needle in the compass box when placed in the magnetic field.

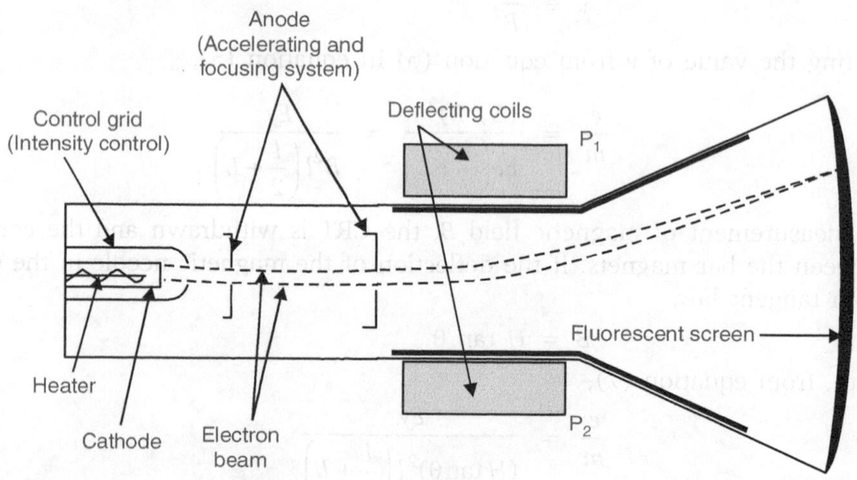

Fig. 24.3. *Cathode ray tube.*

PROCEDURE

1. Place the wooden stand on table such that its arms are in the east-west direction.
2. Mount the CRT in the armed wooden stand such that the tube faces towards north and south.

3. Connect the CRT to the power supply and switch it ON. Adjusting the intensity and focus controls, a fine spot of sufficient brightness is obtained. Focus this spot to the center of the screen by using X and Y shift controls given in power supply (In this position there is no potential applied on the deflecting plates).
4. Now apply a constant potential difference V volt between plates P_1 and P_2 of the CRT with the help of power supply.
5. By doing so, the bright spot on the screen of CRT displaces upwards. Read the displaced position of the spot on the vertical scale.
6. Now take two bar magnets each of uniform intensity and length and place one of them on east arm and other on west arm of the wooden stand with opposite poles facing each other at equal distances from its center.
7. Then move the magnets to and fro such that the displacement of the spot becomes zero *i.e.*, the bright spot comes back to its original position. (Take care that both the magnets are symmetrical relative to the CRT.)
8. Note the distances of the poles of the magnets towards CRT on the scales of the wooden stand say r_1 and r_2.
9. Now remove the magnets from the arms of the wooden stands and reverse the polarity of the voltage V applied to the deflecting plates with the help of power supply.
10. By doing so, the bright spot on the screen of CRT displaces downwards. Read the displaced position of the spot on the vertical scale.
11. Again place the magnets on the arms of the wooden stand and adjust their positions and polarity such that the displacement of the spot becomes zero.
12. Note the distances of the poles of the magnets towards CRT on the scales of the wooden stand say r_1' and r_2'.
13. Switch off the power supply and remove magnets from the wooden stand.
14. Now replace the CRT with compass box such that its center lies exactly on the common axis of the two magnets. Adjust the aluminum pointer of compass box on 0°– 0°.
15. Now place same bar magnets on the arms of wooden stand as they were placed in previous arrangement corresponding to distance r_1 and r_2 respectively and note both ends of the pointer, say θ_1 and θ_2.
16. Then reverse the poles of the bar magnets and place them at distances r_1' and r_2' respectively and again note the two ends of the pointer, say θ_3 and θ_4.
17. Repeat the above steps for different potential differences applying on the deflecting plates.
18. Note the value of horizontal component of the earth's magnetic field H at the experiment place from standard chart.

OBSERVATIONS

❖ For the cathode ray tube

1. Separation between the deflecting plates d = ... mm = ... meter
2. Length of deflection plates l = ... mm = ... meter
3. Distance of screen from the edge of the deflecting plate L = ... mm = ... meter

❖ Horizontal component of earth's magnetic field at ..... H = ... weber/m^2

Table 24.1. For the displacement *y* produced by electric field.

S. No.	Potential difference applied V (volt)	*Potential difference in one direction*			*Potential difference in reverse direction*			*Mean*
		Displacement of the spot on the screen	*Positions of magnets to annul the displacement*		*Displacement of the spot on the screen*	*Positions of magnets to annul the displacement*		*displacement for the potential difference V*
		y_1 (in cm)	*on east arm* r_1 (cm)	*on west arm* r_2 (cm)	y_2 (in cm)	*on east arm* r_1' (cm)	*on west arm* r_2' (cm)	$y = \frac{y_1 + y_2}{2}$ (in cm)

Table 24.2. For the measurement of magnetic field.

S. No.	*When the field is in one direction*				*When the field is in reverse direction*				*Mean deflection* θ
	Positions of magents to annul the displacement		*Deflection of the pointer of compass box*		*Positions of magnets to annul the displacement*		*Deflection of the pointer of compass box*		
	on east arm (r_1) cm	*on west arm* (r_2) cm	*one end* θ_1	*other end* θ_2	*on east arm* (r_1') cm	*on west arm* (r_2') cm	*one end* θ_3	*other end* θ_4	

CALCULATIONS

$$K = \frac{1}{dH^2 l\left(\frac{l}{2}+L\right)} = \text{..................meter/weber}^2$$

From the above observation tables :

S.No.	*Potential difference V (in volt)*	*Mean y (cm)*	*V.y (volt x m)*	*Mean θ (in degree)*	*tan θ*	*tan²θ*	*V.y/tan²θ*

Mean V.y/ $\tan^2\theta$ = ...

Putting the values in the formula, calculate the value of *e/m* for various readings.

$$\frac{e}{m} = K\frac{V.y}{\tan^2\theta} = \text{.................. Coulomb/kg}$$

RESULT

The specific charge of electron = ... Coulomb/kg.

PRECAUTIONS

1. The bright spot on the screen of the cathode ray tube must be strong and well focused.
2. Cathode ray tube should always be placed with its axis along north-south direction.
3. The potential difference applied on the deflecting plates must be such that the vertical displacement of electron beam can be read easily.
4. The magnets should be placed on east-west arm of the wooden stand such that their distance from the cathode ray tube must be symmetrical.

CONCLUSION

VIVAVOCE

1. What do you mean by the specific charge of electron ?
2. What is the principle of your experiment ?
3. What are the requirements to be fulfilled by thermionic emission materials ?
4. What are cathode rays ? Mention their properties.
5. Explain Bethe's law of electron refraction.
6. What is CRT ?
7. Explain electrostatic focusing and deflection of electron beam in CRT.
8. What is the function of an electron gun ?
9. An electron is projected in a magnetic field along the lines of force. How will its motion be affected ?
10. What will happen if an electron is projected in a magnetic field perpendicular to the lines of force ?
11. Explain the motion of electrons in presence of perpendicular electric field.
12. How can we control the electron beam in a CRT ?
13. Is an electron visible to the eye ?
14. How is an electron beam made visible ?
15. What are deflecting plates ? What are their uses ?

❑❑❑

EXPERIMENT—25

AIM

(*i*) To determine the Hall coefficient of a semiconductor.

(*ii*) To determine type of the semiconductor.

(*iii*) To determine concentration of the majority charge carriers.

(*iv*) To determine mobility of the majority charge carriers.

APPARATUS REQUIRED

A Hall effect set up model DHE-21 (Scientific), A Hall probe containing semiconductor wafer, a constant current source, an electromagnet, voltmeter, ammeter, digital gauss meter and probe stand.

Fig. 25.1. *Hall effect experiment setup.*

DESCRIPTION OF THE APPARATUS

The Hall effect set up consists of an electronic digital millivoltmeter and a constant current power supply. The Hall voltage and probe current can be read on the same digital panel meter through a selector switch. A transverse magnetic field is applied with the help of an electromagnet. A gauss meter is used to measure magnetic field. A probe is encapsulated in a non-magnetic sheath, along with the cable through the cutting provided on the rear side of gauss meter. A transport cap is provided for protection of probe. This is mainly used as a transducer for measurement of magnetic field. The Hall probe is a germanium crystal with four spring type pressure contacts is mounted on a sun mica-decorated Bakelite strip. Four leads are provided for connections with measuring devices.

PRINCIPLE

The conductivity experiments cannot answer whether one or more types of carriers either are present or distinguish between them. However, this information can be obtained from Hall effect measurements. When a metal or a semiconductor carrying current I is placed in a transverse magnetic field, an electric field (or potential difference) is produced in the direction perpendicular to both the direction of current and the direction of magnetic field. This phenomenon is known as Hall effect and developed voltage is called as Hall Voltage.

THEORY

Consider a slab of a specimen of p-type semiconductor having thickness t and area of cross-section A. When an external electric field E_x is applied along its axis (*i.e.*, X-axis), a current I flows. The holes will drift with velocity v_x in direction of the electric field (Fig. 25.2).

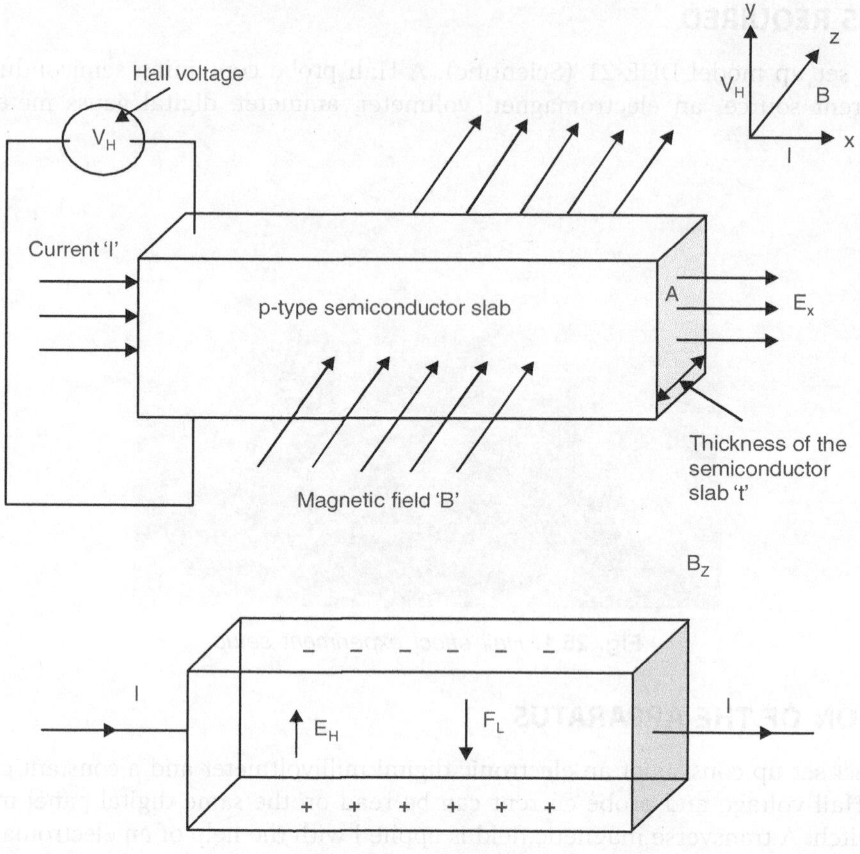

Fig. 25.2. *Hall effect.*

Because the current is carried by the holes (majority carriers), their velocity v_x will be in the same direction as that of current.

Now if a transverse magnetic field B_z is applied along Z-axis, it will exert a transverse magnetic deflection force (Lorentz Force) in downward direction (–ve Y-axis)

$$(F_L)_y = ev_x B_Z \qquad ...(1)$$

Which tends to deflect charge carriers in the semiconductor to drift downward to one of its face, leading to an accumulation of charges in that face.

Consequently lower surface becomes positive and upper surface becomes negative. As a result an electric field E_H (Hall field) is produced in upward direction (+ve Y-axis) between the two faces, which opposes the Lorentz force due to magnetic field and hence opposes transverse drifting of holes.

The force F_E due to electric field is given by

$$F_E = eE_H \quad \text{...(2)}$$

The electric field builds till it exactly compensates the effect of magnetic field. At equilibrium the net force on the holes becomes zero *i.e.,*

$$F_E = F_L$$

$$eE_H = ev_x B_Z$$

Thus, $$E_H = v_x B_Z \quad \text{...(3)}$$

Let the number of holes (or charge carriers) per unit volume in the slab is p, then current density $J = pev_x$

or $$v_x = \frac{J}{pe}$$

Then from equation (3)

$$E_H = \frac{1}{pe}(J.B_Z)$$

or $$E_H = R_{Hall}(J.B_Z) \quad \text{...(4)}$$

where, $$R_{Hall} = \frac{1}{pe} = \textbf{Hall coefficient} \quad \text{...(5)}$$

If b is the width of the slab and t is its thickness then current density in the slab along x-axis is $J = \frac{I}{b.t}$, then from equation (4),

$$E_H = R_{Hall}\frac{IB_Z}{b.t}$$

And, Hall voltage $V_H = E_H \times b = R_{Hall}\frac{IB_Z}{t}$

Thus, $$R_{Hall} = \frac{V_H t}{IB_Z} \quad \text{...(6)}$$

So by knowing the thickness t of the semiconductor, the magnetic field B_Z and by measuring the Hall voltage V_H for the given current I, the Hall coefficient R_{Hall} can be calculated from equation (6). For positive value of R_{Hall}, given semiconductor is of p-type and for negative value of R_{Hall}, the given semiconductor is of n-type.

Similarly by knowing Hall coefficient, the concentration of the charge carriers can be determined from equation (5), *i.e.,*

$$p = \frac{1}{eR_{Hall}} \quad \text{...(7)}$$

Also the mobility of the charge carriers μ in the material can be obtained from the following expression

$$\mu = \frac{R_{Hall}}{\rho} \qquad [\because \sigma = ne\mu;\ \mu = \frac{1}{ne\sigma}] \quad \text{...(8)}$$

FORMULA USED

1. The Hall coefficient

$$R_{Hall} = \frac{V_H t}{IB_Z}$$

2. The concentration of the charge carriers (holes)

$$p = \frac{1}{eR_{Hall}}$$

3. The mobility of the carriers

$$\mu = \frac{R_{Hall}}{\rho}$$

where R_{Hall} = Hall coefficient.

V_H = Hall voltage.

B = Magnetic field.

I = Current flowing through semiconductor.

$p = n$ = Concentration of charge carriers (Holes or electrons).

e = Charge of the electron.

μ = Mobility of the charge carriers.

ρ = Resistivity of the semiconductor.

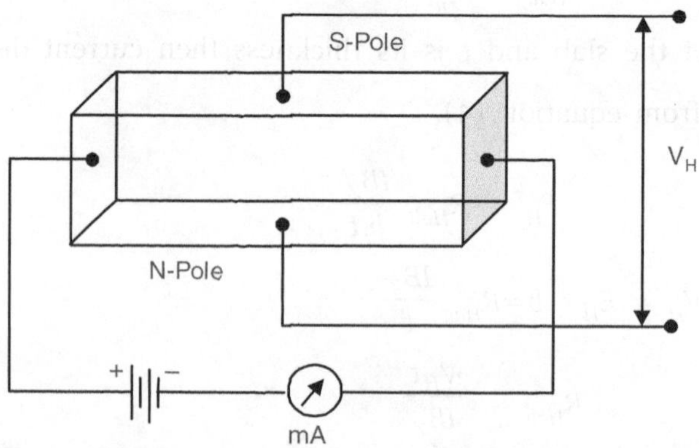

Fig. 25.3. *Experimental arrangement for the study of Hall effect.*

PROCEDURE

1. Switch on the power supply and adjust the magnetic field to a suitable value by using digital gauss meter. The reading of gauss meter is adjusted to zero with 'ZERO' control with the probe away from any magnetic field. The probe cap is removed and the probe is kept in the magnetic field to be measured. The flat face of the probe is kept perpendicular to the direction of the magnetic field. The reading of the meter multiplied by the range value gives the flux density of the magnetic field in gauss. If the field strength is not known it is advisable to start with x10 range.
2. Note this reading of magnetic field. If the magnetic field indicated by the gauss meter is positive (without sign), the pole facing the side of the Gauss meter probe marked 'N' is 'North pole'.

3. Next mount the semiconductor probe on the probe stand carefully.
4. Now connect the widthwise contacts of the Hall probe to the voltage terminals of the Hall effect setup.
5. Similarly connect the lengthwise contacts of the Hall probe to the current terminals of the Hall effect setup.
6. Now place the Hall probe in the air gap of the electromagnet.
7. Now switch on Hall effect setup and rotate the Hall probe in a vertical plane till the Hall voltage generation is a maximum. This is the position when direction of current in the probe and magnetic field would be perpendicular to each other.
8. Adjust the current through the semiconductor to a suitable value. For this current note down the Hall voltage and its polarity.
9. Now repeat the above process for different values of current. Note down all the observations in the table.

OBSERVATIONS

Thickness of the semiconductor wafer 't' = ... mm

= .... cm

Resistivity 'ρ' = ... ohm.cm.

Magnetic field, 'B' = ... gauss = ... Tesla

Table 25.1

S.No.	*Current through the semiconductor I (mA)*	*Hall voltage V_H (mA)*

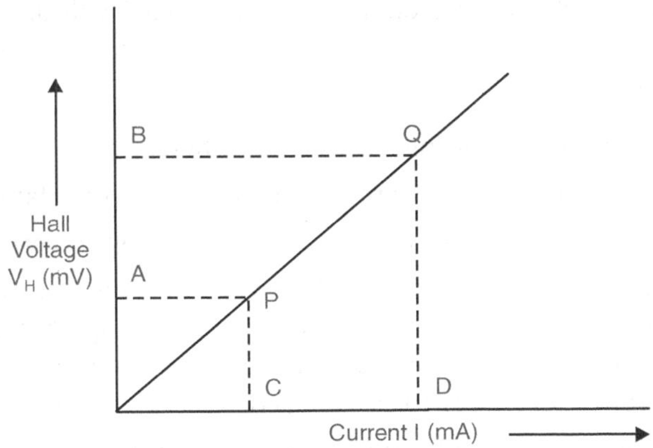

Fig. 25.4. *Graph.*

CALCULATIONS

From the above observation table, plot a graph by taking the current I on X-axis and the Hall voltage V_H on Y-axis. We get a straight line as shown in Fig. 25.4.

$$\text{The slope of the line } s = \frac{V_H}{I} = \frac{AB}{CD} = \ldots\ldots\ldots\ldots\ldots \text{ V/A}$$

1. Hall coefficient

$$R_{Hall} = \frac{V_H t}{I B_Z} = \frac{t}{B_Z} \times \text{Slope of } V_H - I \text{ graph}$$

$$= \frac{t}{B_Z} \times s = \ldots\ldots\ldots\ldots \text{cm}^3\text{/Coulomb}$$

2. Majority charge carriers concentration

$$p\text{(or)}n = \frac{1}{eR_{Hall}} = \ldots\ldots\ldots\ldots\ldots\ldots\ldots \text{/cm}^3$$

3. Mobility of the charge carriers

$$\mu = \frac{R_{Hall}}{\rho} = \ldots\ldots\ldots\ldots\ldots \text{ cm}^2\text{/volt s.}$$

RESULT

1. The calculated value of Hall coefficient R_{Hall} = ... cm³/Coulomb.
2. The sign of the Hall coefficient is ....
   Therefore the given semiconductor is of .... type.
3. The concentration of the charge carriers p (or n) = ... /cm³.
4. The mobility of the charge carriers μ = ... cm²/volt s.

PRECAUTIONS

1. Mount probe properly for getting maximum voltage.
2. Vary magnetic field and current carefully in steps to avoid damage of semiconductor.
3. Handle probes carefully and do not touch probe during experiment.
4. It is advisable that the probe should not be used at temperature higher than 50°C. The probe is very delicate and hence it is advised that its cap should be inserted whenever the probe is not in use and kept inside the instrument along with the cable. Any kind of strain on the probe element is to be avoided.
5. The current through the sample should not be large enough to cause heating.

CONCLUSION

VIVAVOCE

1. How many semiconductors are there in the periodic table ?
2. What is the resistivity of semiconductors at room temperature ?
3. How semiconductor acts at 0 K ?
4. What are *p*-type and *n*-type semiconductors ?
5. What are donors and accepters ?
6. Why do we prefer extrinsic semiconductors than intrinsic ?
7. What is the meaning of band gap of a semiconductor ?
8. What is Hall effect ? What is its use in the semiconductor ?
9. What is the effect of doped impurities and increase of temperature in a semiconductor ?
10. What are diffusion and drift current in a semiconductor ?
11. Define minority carriers' lifetime.
12. What is Fermi level in a semiconductor ?
13. What are the applications of Hall effect ?
14. What do you mean by Lorentz force ?
15. How can you determine the direction of Lorentz force ?
16. What do you mean by the mobility of charge carriers ?
17. What is drifting of charge carriers ? Explain.
18. Define Hall coefficient.
19. On what parameters does the Lorentz force due to magnetic field depend ? Give expression for it.
20. How can you determine whether the given semiconductor is *p*-type or *n*-type with the help of Hall effect experiment ?

❑❑❑

Experiment—26

AIM

(*i*) To study the current-voltage characteristics of photo-resistor at constant light intensity ϕ.

(*ii*) To study the current-intensity characteristics of photo-resistor at constant voltage V.

APPARATUS REQUIRED

Photo-resistor, lamp, optical bench, adjustable slit, polarization filters, lens, power supply and milliammeter.

Fig. 26.1. *Photo-resistor kit.*

DESCRIPTION OF THE APPARATUS

A photo-resistor is a large surface slab of a semiconductor in bulk form or a thin film of semiconductor deposited on an insulating substrate with ohmic contacts at opposite ends. The semiconductor materials most commonly used are cadmium compounds. In the experiment, a CdS photo-resistor is exposed to light from a lamp. When radiation falls on it, its conductivity increases *i.e.,* a resistance decrease, that's why it is called "photo-resistor".

Optical riders are mounted on the optical bench. A photo-resistor is attached at the left edge of the optical bench. The lamp with a spherical condenser is placed at the right edge of the optical bench and connected with a power supply (6 V–12 V). A lens is used to focus the ray of light on the photo-resistor. Two polarization filters and an adjustable slit are mounted between lamp and lens. The intensity of light can be varied with different polarization angle. The positive terminal of a D.C. power supply is connected with positive terminal of the photo-resistor and the negative terminal of D.C. power supply is connected with positive terminal of milliammeter. The other end of the milliammeter is connected with negative terminal of photo-resistor. A voltmeter is connected parallel to the photo-resistor to measure potential across photo-resistor. The photocurrent I_{ph} is measured with milliammeter connected in series.

PRINCIPLE

When radiation of suitable frequency falls upon a semiconductor, its conductivity increases. This phenomenon is called "Photoconductive effect", and is the basis of a photoconductor. The conductivity of a material is proportional to the concentration of charge carriers in it. In a semiconductor, these are thermally generated electron-hole pairs. When radiation falls on it, covalent bonds are broken and additional pairs are created. These increased charge carriers increase the conductivity or decrease the resistance of the semiconductor.

In the absence of radiation, on applying a potential difference Fig. 26.2, a small circuit current flows, which is called dark current. When radiation falls on the photo-resistor additional charge carriers generated. These charge carriers move under the applied potential difference across the photoconductor, reach the ohmic contacts at the ends of the semiconductor and constitute the photocurrent which is added to the dark current.

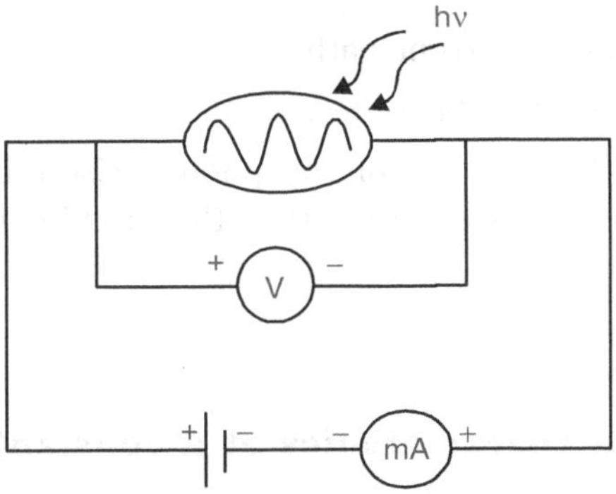

Fig. 26.2. *Circuit diagram.*

THEORY

The energy level diagram of a semiconductor is shown in Fig. 26.3. Let the energy gap between valence band and conduction band is E_g. If the frequency of the incident photon ν is so low that $h\nu < E_g$, light passes through the material with very little absorption. In terms of energy band diagram of intrinsic semiconductor, this energy $h\nu$ is inadequate in transferring an electron from valence band into conduction band. But if $h\nu > E_g$, an electron absorbs a photon from the incident radiation and jumps from valence band into the conduction band and new electron-hole pair is generated, in addition to thermal electrons. The electrons and holes serve as the carriers of electrical conductivity. The electrons are in the conduction band and holes are in the valence band of the semiconductor. This results in increased conductivity. Hence such a material is called photo-conductor. Thus in photoconductivity, the conductivity of a semiconductor is raised by striking radiation on the sample.

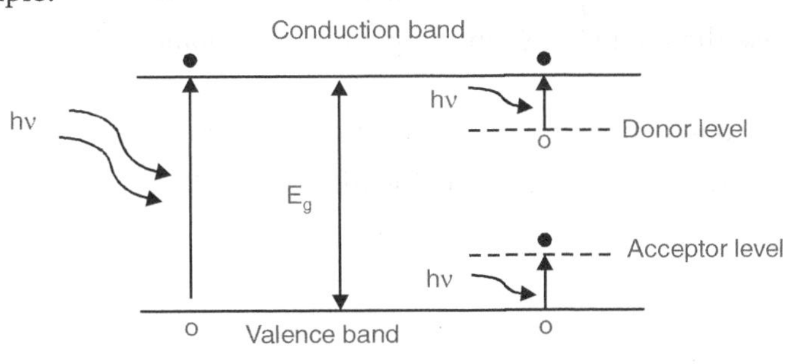

Fig. 26.3. *Energy level diagram.*

The increase in conductivity is given by

$$\Delta \sigma = \Delta p e \mu_p + \Delta p e \mu_n \quad ...(1)$$

where, e = Elementary charge.

Δp = Change of the hole concentration.

Δn = Change of the electron concentration.

μ_p = Mobility of the holes.

μ_n = Mobility of the electrons.

When a voltage V is applied, the photocurrent is

$$I_{ph} = \frac{\Delta\sigma . V . A}{d} \quad [\text{as } I = V/R] \quad ...(2)$$

where A = Cross-section of the current path.

d = Distance between the electrodes.

The photocurrent I_{ph} at the position of the photo-resistor being varied by means of two polarization filters which are placed one behind the other and by varying positive potential at the photo-resistor.

PROCEDURE

A. For current–voltage characteristics at a constant light intensity ϕ.

1. Illuminate the lamp, so that the maximum homogeneous light rays fall on the photo-resistor.
2. Set an angle $\alpha = 0°$ between the polarization planes of the filters.
3. Switch on the power supply and set potential at an appropriate value (say 18 V).
4. Measure photocurrent I_{ph} with the help of milliammeter.
5. Now reduce the voltage V to 0 V in steps of 2 V. Measure the photocurrent I_{ph}, each time and record it.
6. Repeat the series of measurement for different polarization angle, $\alpha = 30°, 60°$.

B. For current–irradiance characteristics at a constant voltage V.

1. Switch on the power supply and set potential at an appropriate value (say 18 V).
2. Set an angle $\alpha = 10°$ between the polarization planes of the filters.
3. Measure photocurrent I_{ph} with the help of milliammeter.
4. Now increase the angle between the polarization planes of the filters up to 90° in steps of 10°.
5. Measure the photocurrent I_{ph} , each time and record it.
6. Repeat the series of measurement for different voltages.

OBSERVATION

Table 26.1. To study the current-voltage characteristics of photo-resistor at a constant light intensity ϕ.

Voltage V (in Volts)	I_{ph} *(0°) (mA)*	I_{ph} *(30°) (mA)*	I_{ph} *(60°) (mA)*
18			
16			
14			
12			
10			
8			
6			
4			
2			

Table 26.2. To study the current-intensity characteristics of photo-resistor at a constant voltage V.

α	*For V = ...*	*For V = ...*	*For V = ...*
	I_{ph} *(mA)*	I_{ph} *(mA)*	I_{ph} *(mA)*
0°			
10°			
20°			
30°			
40°			
50°			
60°			
70°			
80°			
90°			

GRAPHICAL EVALUATION

The relation between the photocurrent I_{ph} and the voltage V applied at a constant intensity (constant angle α between the polarization planes of the filters), *i.e.*, the current-voltage characteristics, is shown in Fig. 26.4. Even at an angle $\alpha = 90°$ between the polarization planes of the filters a photocurrent flows, since in this position the polarization filters do not extinguish the ray of light completely.

It is clear from the graph the slope of each characteristic depends on the irradiance.

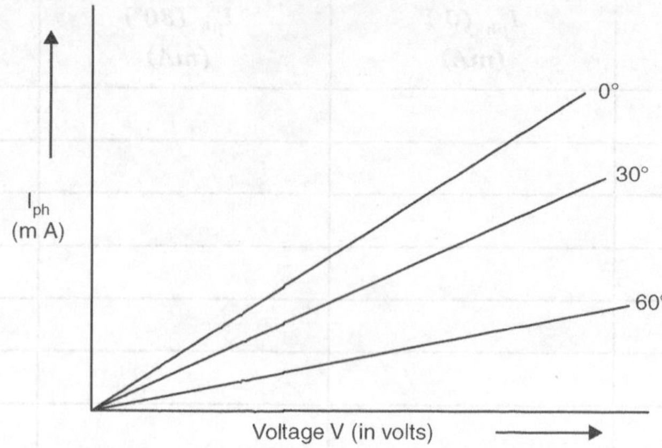

Fig. 26.4

To study the current–intensity characteristics of photo-resistor at a constant voltage V

The graph between the photocurrent I_{ph} and the irradiance at a constant voltage, the current-intensity characteristics is shown in Fig. 26.5.

It is clear from the graph, the photocurrent increases with increasing irradiance. However, the characteristics are not perfectly linear. The slope rather decreases with increasing intensity.

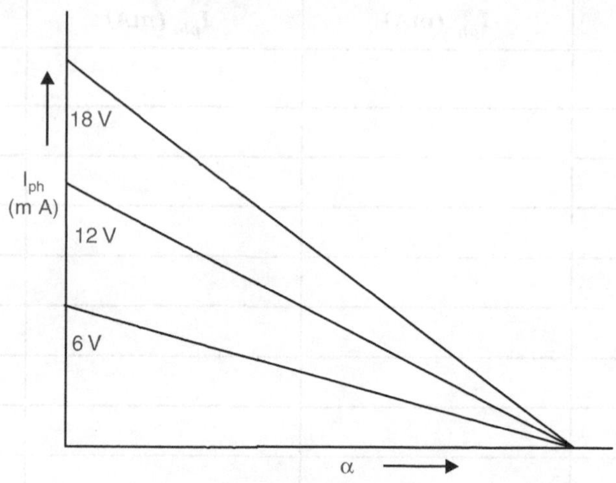

Fig. 26.5

RESULTS

The CdS photo-resistor behaves like an ohmic resistance that depends on the intensity of light.

PRECAUTIONS

1. The photo-resistor is influenced even by slight residual light in the experiment room. Therefore darken the experiment room so that the measuring instruments can just be read.
2. When the illumination is changed, take readings after a while until the variation in photocurrent is stopped.
3. Adjust the slit so that light falls homogeneously on photo-resistor.

CONCLUSION

VIVAVOCE

1. What is photoconductivity ?

2. What are photoelectrons ?

3. Explain the meaning of threshold wavelength in photoelectric effect.

4. Explain retarding potential.

5. Define work function.

6. An increase in frequency of the incident light increases the velocity of the ejected photoelectrons, why ?

7. Which photon is more energetic violet or red one ?

8. What is photoelectric cell ?

9. Mention the types of photoelectric cells.

10. What effect will be done on photocurrent of varying potential, intensity of light and frequency of incident light ?

❑❑❑

EXPERIMENT—27

AIM

To determine the value of Planck's constant (h) with the help of a photocell.

APPARATUS REQUIRED

A photocell, a battery, a micro ammeter, a rheostat, voltmeter, light source (mercury lamp), convex lens, optical filters of different wavelength.

DESCRIPTION OF THE APPARATUS

Photocell is a device, which converts light energy into electrical energy. It consists of an evacuated glass or quartz tube. Inside the tube there is a metal surface coated with photosensitive material C and a wire loop A, which act as cathode and anode respectively. The cathode and anode are connected to metallic pins and through these pins; photocell is connected to the external circuit.

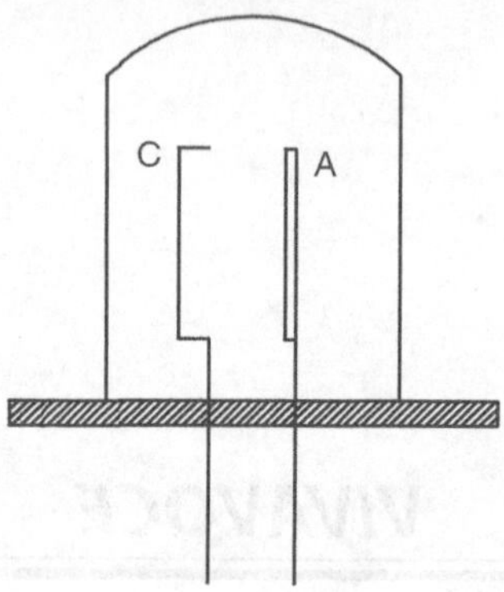

Fig. 27.1. *Photoemissive cell.*

The experimental arrangement is shown in Fig. 27.2. Light from the source (mercury lamp) is made incident on the photocell placed inside a wooden box after passing through the filter by means of a convex lens. The anode of the photocell is connected with positive terminal of the power supply (battery). Rheostat is used as a potential divider. A voltmeter measures the voltage applied to the anode of photocell and a micro-ammeter measures (μA) the photoelectric current.

PRINCIPLE

When light of a suitable frequency is allowed to fall on the cathode of cell, photoelectrons emit. These electrons are attracted by anode, which is kept at positive potential with respect to cathode. So a current starts flowing in the circuit. Due to it micro-ammeter shows a deflection. If a negative potential is applied on the anode photocurrent decreases and it becomes zero at a certain negative potential. The minimum negative potential, at which photocurrent becomes zero, is called stopping potential or cut-off potential. For the given frequency of incident radiation, the stopping potential is independent of its intensity. But it depends on the wavelength of incident light, lower the wavelength of incident light, higher is the stopping potential and vice versa.

THEORY

For a given frequency of the incident radiation, the stopping potential V_0 is related to the maximum kinetic energy of the photoelectron ejected.

If m is the mass and v_{max} is the maximum velocity of photoelectron ejected, then maximum kinetic energy of the electron

$$E_k = \frac{1}{2}mv^2_{max}$$

If e is the charge of the electron and V_0 is the stopping potential, then work done by the retarding potential in stopping the electron $= eV_0$

Therefore, we have

$$\frac{1}{2}mv^2_{max} = eV_0 \quad \text{...(1)}$$

Threshold energy or work function is the minimum energy required to eject electron from the photosensitive surface. According to Einstein when light of wavelength λ (frequency ν) is made incident on a photosensitive surface photons of energy $\frac{hc}{\lambda}$ (= $h\nu$) fall on the surface. The energy of the photon is spent in two ways :

(*a*) A part of the energy of the photon is used in ejecting the electron from surface that is equal to the work function W_0.

(*b*) The rest of the energy of the photon is used in imparting the kinetic energy to the emitted photoelectrons.

If v is the velocity of the emitted photoelectron and m is its mass, then

$$\text{Kinetic Energy of the photoelectron} = \frac{1}{2}mv^2$$

Therefore,

$$\frac{hc}{\lambda} = \frac{1}{2}mv^2 + W_0 \quad \text{...(2)}$$

or

$$\frac{1}{2}mv^2 = \frac{hc}{\lambda} - W_0 \quad \text{...(3)}$$

Now if V_0 is the stopping potential, then

$$eV_0 = \frac{1}{2}mv^2 \quad \text{...(4)}$$

Therefore from equations (3) and (4)

$$eV_0 = \frac{hc}{\lambda} - W_0 \quad \text{...(5)}$$

It is clear from equation (5) that stopping potential V_0 is inversely proportional to the wavelength λ. If a graph is plotted for V_0 versus $\frac{1}{\lambda}$, it will be a straight line with slope $s = \frac{hc}{e}$. Therefore finding the slope s, the value of Planck's constant h can be determined.

FORMULA USED

Planck's constant

$$h = \frac{es}{c}$$

$$h = \frac{e}{c} \times (\text{slope of graph plotted for } V_0 \text{ versus } \frac{1}{\lambda})$$

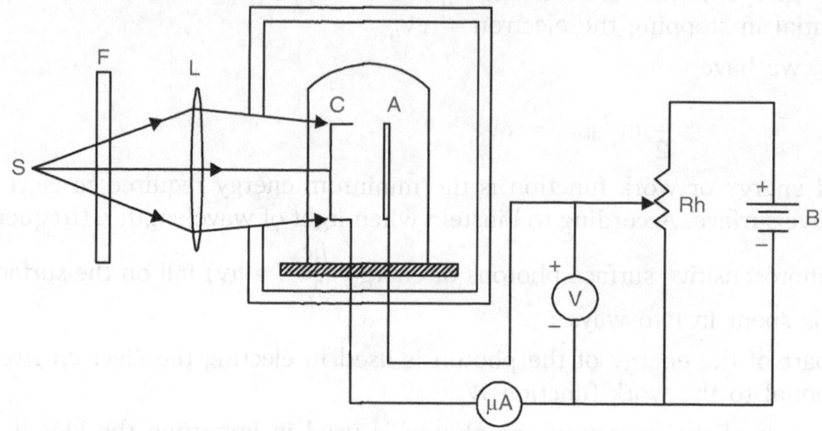

Fig. 27.2. *Experimental arrangement.*

PROCEDURE

1. Make experimental arrangement as shown in Fig. 27.2.
2. Light from the lamp *S* after passing through the filter *F* is made incident on the photocell placed inside a wooden box by means of the convex lens.
3. Keep the anode of the photocell at zero potential. The potential can be measure with voltmeter *V*.
4. Taking the red filter of wavelength and note the photocurrent with the help of microammeter (μA) corresponding to zero potential.
5. Now vary the anode potential in steps of 0.2 volt (*i.e.*, –0.2 V, –0.4 V, ...) and record the photoelectric current corresponding to each anode potential, till it becomes zero.
6. Repeat the above steps (4) and (5) for green and yellow filters of wavelength.
7. From observation table, plot a graph between current '*I*' on *Y*-axis and voltage '*V*' on *X*-axis. The graph will be in the form of three curves (Fig. 27.3).
8. Note the values of intercepts on *X*-axis corresponding to each wavelength, which gives the stopping potentials V_0 corresponding to those wavelengths.
9. Now, plot a graph between stopping potential V_0 on *Y*-axis and $\frac{1}{\lambda}$ on *X*-axis (Fig. 27.4).

OBSERVATIONS

Table 27.1

S. No.	*For filter 1 (wavelength λ = ...)*		*For filter 2 (wavelength λ = ...)*		*For filter 3 (wavelength λ = ...)*		*For filter 4 (wavelength λ = ...)*		*For filter 5 (wavelength λ = ...)*	
	Negative potential on anode 'V' (Volt)	*Current 'I' (μA)*	*Negative potential on anode 'V' (Volt)*	*Current 'I' (μA)*	*Negative potential on anode 'V' (Volt)*	*Current 'I' (μA)*	*Negative potential on anode 'V' (Volt)*	*Current 'I' (μA)*	*Negative potential on anode 'V' (Volt)*	*Current 'I' (μA)*
1.										
2.										
3.										
4.										
5.										
6.										
7.										
8.										
9.										
10.										

CALCULATIONS AND GRAPH

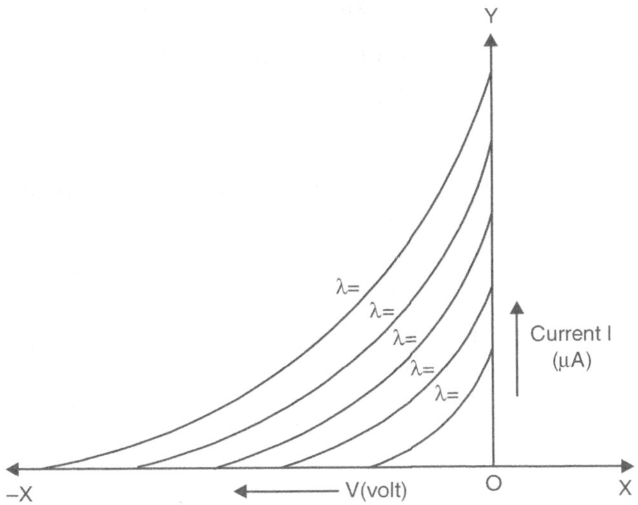

Fig. 27.3

Fig. 27.3 is the graph between potential and current. Now, corresponding to each wavelength the intercepts are noted which represents stopping potential. Now tabulate the stopping potentials for each wavelength.

Table 27.2

S.No.	Wavelength of incident light 'λ' (meter)	Stopping potential 'V_0' (volt)	$\frac{1}{\lambda}$

Now plot a graph between stopping potential V_0 and $\frac{1}{\lambda}$.

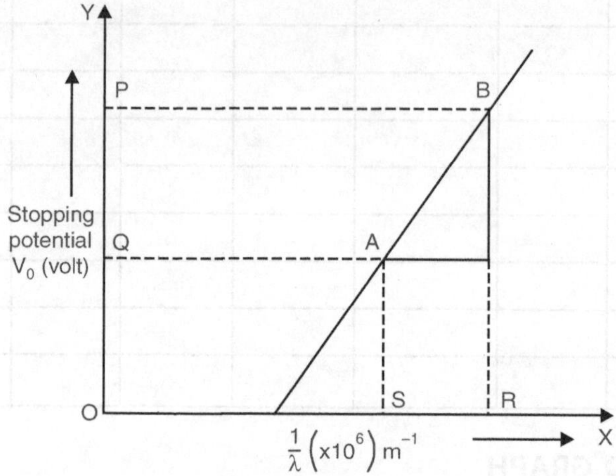

Fig. 27.4

The graph is straight line. The slope of the straight line

$$s = \frac{PQ}{RS} = \text{.............} \times 10^6$$

From equation $eV_0 = \frac{hc}{\lambda} - W_0$, the slope of V_0, versus $\frac{1}{\lambda}$ curve is

$$s = \frac{hc}{e}$$

Hence, Planck's constant $h = \frac{e}{c} \times$ (slope of graph plotted for V_0 versus $\frac{1}{\lambda}$)

$$= \text{....................} \frac{C}{m/s} \times \frac{V}{1/m} = \text{Js}$$

RESULT

It is clear from Fig. 27.4 that the stopping potential decreases with the increase in the wavelength of incident light.

Planck's constant = ... joule second

Standard value of $h = 6.6 \times 10^{-34}$ joule second.

PRECAUTIONS

1. The light must fall on the entire surface of cathode of photocell.
2. Extra light must not fall on the photocell, hence the experiment should be performed in a dark room.
3. The intensity of light must be high.

CONCLUSION

VIVAVOCE

1. What is the object of your experiment ?
2. What is a photocell ?
3. How many types of photocell do you know ?
4. Which type of photocell is being used in this experiment ?
5. What is photoelectric effect ?
6. Explain Einstein's equation in photoelectric effect.
7. How do you measure the kinetic energy of photoelectron in your experiment ?
8. What is work function ?
9. Can the work function depend on wavelength of the incident light ?
10. What do you mean by stopping potential ?
11. How does the stopping potential depend upon the wavelength of incident light ?
12. Does stopping potential depend upon the intensity of the incident light ?
13. What is the value of Planck's constant ?
14. What is the unit of Planck's constant ?
15. Explain the dependence of intensity of light and frequency of light on stopping potentials.
16. What do you mean by intensity and frequency of light according to quantum theory ?
17. What do you mean by negative potential ?
18. What do you mean by work function or threshold energy of a metal surface ?
19. What happen if we increase the frequency or wavelength of incident radiation ?
20. What is stopping potential ? Why it is negative ?
21. What are the applications of this effect ?

❑❑❑

EXPERIMENT—28

AIM

To study characteristic curves of the given *p-n* junction diode.

APPARATUS REQUIRED

A *p-n* junction diode, a regulated power supply with 0 to 12 volt output voltage, a milliammeter, a microammeter, a voltmeter, a rheostat or potentiometer and connecting wires.

Fig. 28.1. *p-n Junction diode characteristic apparatus.*

DESCRIPTION

Apparatus

A *p-n* junction diode experimental kit is shown in Fig. 28.1. Two D.C. power supplies used here for forward and reverse biasing. With the help of one rheostat (or variable resistance) we can vary the forward bias voltage and reverse bias voltage is varied with the help of other rheostat. These voltages are read by the two voltmeters. Similarly current in the forward biasing is read by the milliammeter connected in series and reverse biasing current is read by the microammeter connected in series.

p-n Junction Diode

It is a single piece of semiconductor crystal having an excess of donor impurities (*n*-type) into one side and of acceptor impurities (*p*-type) into the other. By a special manufacturing process, *p* and *n*-type semiconductors can be melted so that a boundary or junction is formed between them. This junction is extremely thin and of the order of 10^{-3} mm. It is called a *p-n* junction (the lattice structure remaining continues).

The semiconductor containing a *p-n* junction is called semiconductor diode. A *p-n* junction diode is represented by an arrow and bar. In diode symbol in Fig. 28.2, the low resistance is from left to right (towards the triangle point) and the high resistance is the opposite direction.

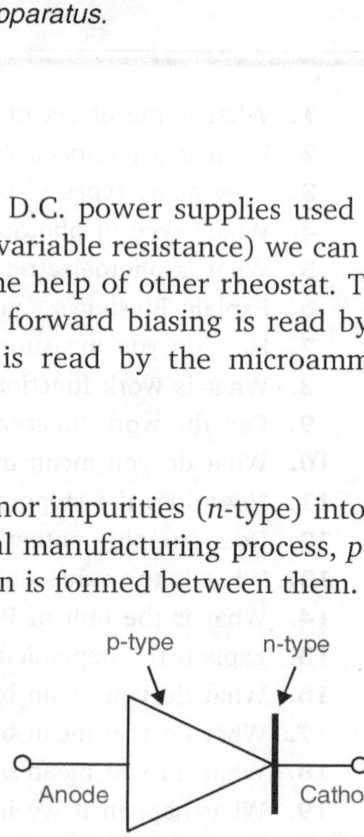

Fig. 28.2. *Circuit symbol.*

When diode is connected to an external D.C. voltage source, then the diode is said to be biased. Milliammeter and micro-ammeter are used for measuring current in the biasing condition.

PRINCIPLE

Semiconductors, pure or doped (*p*-type or *n*-type), are bilateral; current flows in either direction with equal ease. If however, in a semiconductor there exist a *p*-type region on one side and *n*-type region on the other, then the semiconductor becomes unilateral, current flows easily in one direction only.

THEORY

The energy level diagram of the *p-n* junction may be visualised most easily by considering the *p*-type and *n*-type semiconductors as being initially separated.

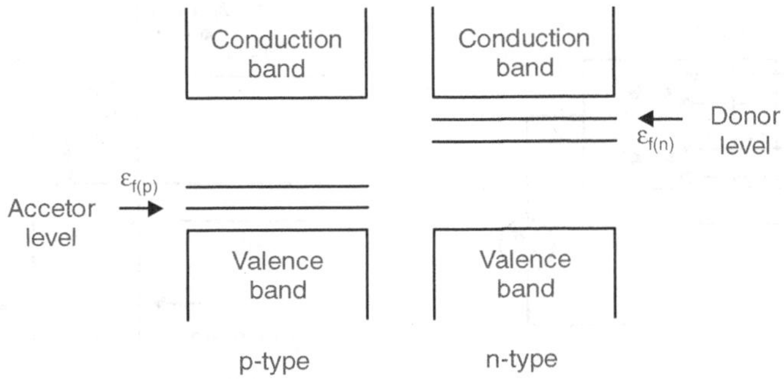

Fig. 28.3. *Energy level diagram.*

The Fermi energy ε_f (highest energy of the filled electron states) is closer to the top of the valence band in the *p*-type material and closer to the bottom of the conduction band in the *n*-type. When a *p-n* junction is first created, conduction band (mobile) electrons from the *n*-doped region diffuse into the *p*-doped region where there is a large population of holes with which the electrons "recombine". When a mobile electron recombines with a hole, the hole disappear and an excess negative charge appear on this side (ionised atoms).

In a similar way an excess positive charge builds-up in the *n*-side, when holes diffuse across the junction. Thus, two mobile charge carriers have vanished near junction. The region around the

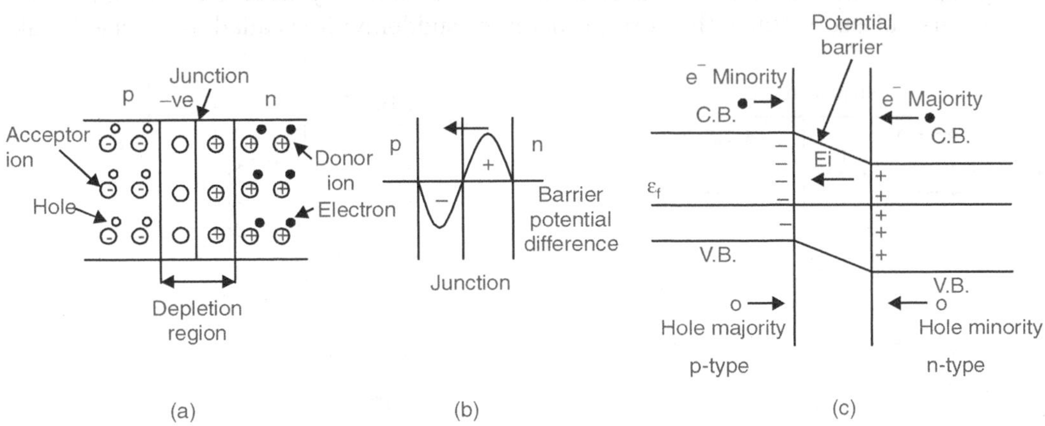

Fig. 28.4. *Formation of depletion region.*

p-n junction becomes depleted of free charge carriers and thus behaves as an insulator. This region is called as Depletion region. This region behaves like a barrier; it prevents further flow of majority charge carriers but allows flow of minority charge carriers. When the flow ceases an emf or potential difference is produced and it has a magnitude of few tenth of a volt.

A *p-n* junction diode can be biased in two ways :

1. If the positive terminal of the battery is connected to *p*-region and the negative terminal is connected to *n*-region, the diode is said to be forward biased. In this biasing with an emf greater than the barrier potential difference the electrons of *n*-region are given extra energy by negative terminal of the battery and move towards the junction. Similarly the holes of *p*-region are forced by positive terminal of the battery and move towards the junction. Thus conduction takes place at the junction. By increasing the potential difference across the diode, the current in diode also increases.

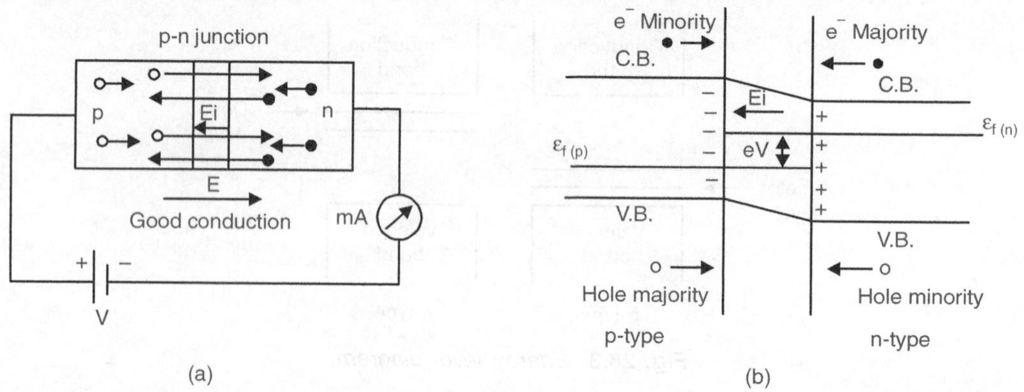

Fig. 28.5. *Forward bias.*

2. If the positive terminal of the battery is connected to *n*-region and the negative terminal is connected to *p*-region, the diode is said to be reverse biased. In this biasing the electrons of *n*-region are attracted by positive terminal of the battery and move away from the junction. Similarly the holes of *p*-region are attracted by negative terminal of the battery and move away from the junction. Thus no current flows due to majority charge carriers. But the minority charge carriers viz. holes of *n*-region and electrons of *p*-region move towards the junction due to which a very weak current flows. At a high reverse biased voltage, the current increases suddenly; it is called as reverse breakdown.

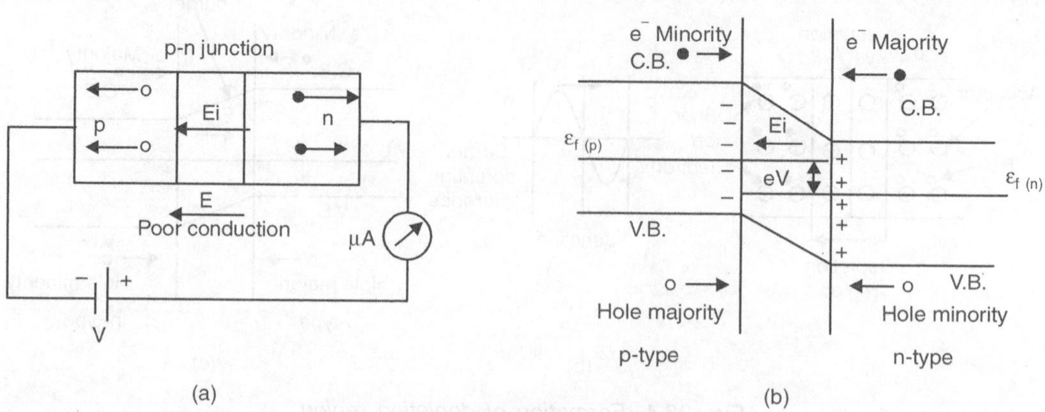

Fig. 28.6. *Reverse bias.*

CIRCUIT DIAGRAM

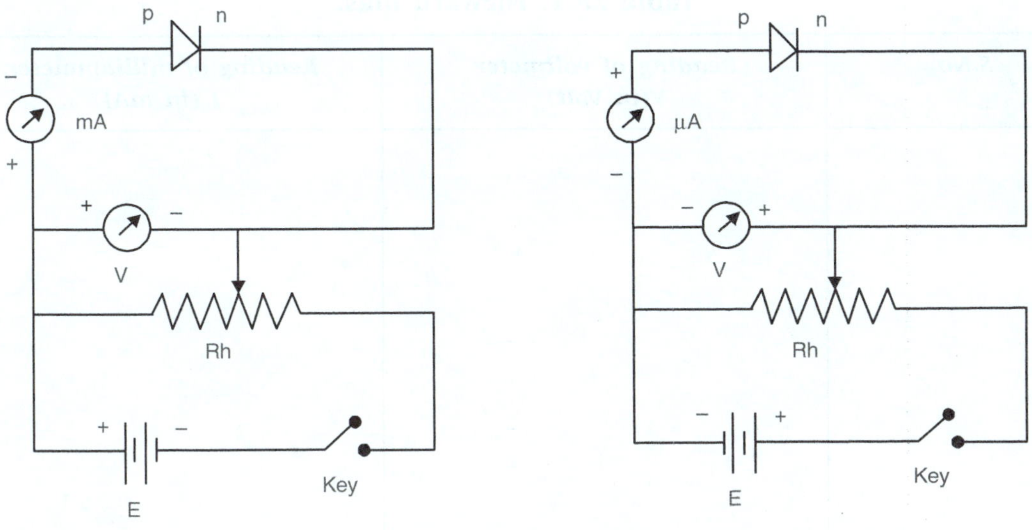

Fig. 28.7. *Circuit diagram.*

PROCEDURE

A. In forward bias :

1. Connect the diode and other instruments as shown in Fig. 28.7 (*a*). Connect *P* terminal of the diode to the positive terminal of the battery and *N* terminal of the diode to the negative terminal of the battery. To read the current a milliammeter (mA) and to read the voltage a voltmeter (V) are used as shown in Fig. 28.7 (*a*).
2. First switch on the power supply.
3. Set the voltage at 0 V and record the current through the diode.
4. Now increase the value of voltage with the help of rheostat or potentiometer in steps of 1 V. For each value of voltage record the corresponding current value from milliammeter. Note all the observations in Table 28.1.

B. In reverse bias :

1. For reverse biasing connect the circuit as shown in Fig. 28.7 (*b*). Connect *P* terminal of the diode to the negative terminal of the battery and *N* terminal of the diode to the positive terminal of the battery. To read the current microammeter is used.
2. First switch on the power supply.
3. Set the voltage at 0 V and record the current through the diode.
4. Now increase the value of voltage with the help of rheostat or potentiometer in steps of 1 V. For each value of voltage record the corresponding current value from micro-ammeter. Note all the observations in Table 28.2.

OBSERVATIONS

Table 28.1. Forward bias.

S.No.	Reading of voltmeter V(in Volt)	Reading of milliammeter I (in mA)

Table 28.2. Reverse bias.

S.No.	Reading of voltmeter V (in Volt)	Reading of micro-ammeter I (in μA)

GRAPH

Plot the volt-ampere graph. In forward bias take voltage (*V*) on the *X*-axis and current (mA) on the *Y*-axis. In case of reverse bias take voltage (*V*) on the (–*X*)-axis and current (μA) on the (–*Y*)-axis. The characteristic curve is shown in Fig. 28.8.

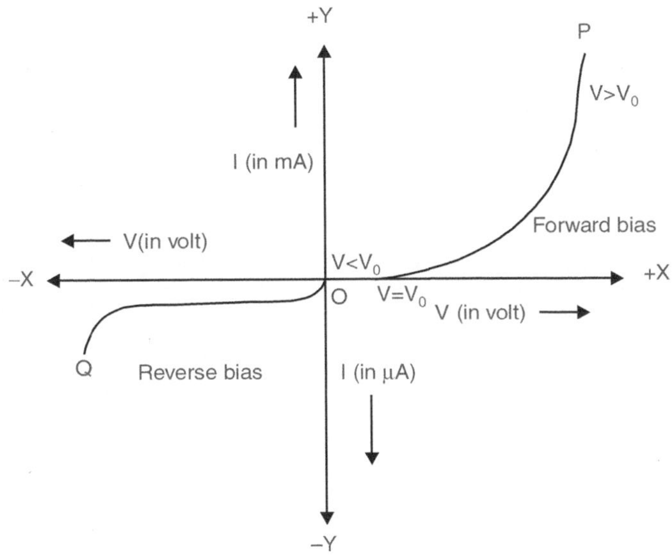

Fig. 28.8. *Characteristic curve.*

RESULT

The characteristics I-V curve *POQ* (in Fig. 28.8) of the *p-n* junction shows that the *p-n* junction acts as a rectifier. It has a low resistance in one direction of potential difference (+ve) and a high resistance in the opposite polarity (–ve). It is clear that in forward bias the current is more due to flow of the majority charge carriers. Order of the current is mA. As the voltage is increased, the current also increases. Initially for the values of applied forward voltage (V) less than the barrier potential (V_0) no current starts flowing but as soon as V reaches the value same as V_0 current starts flowing. For $V > V_0$ current almost becomes linear as in the case of a simple resistance wire.

But in reverse bias, the current is very low (of the order of μA) shows that the current is mainly due to the flow of the minority charge carriers. Since the minority carriers are thermally generated, the magnitude of the reverse current depends on the temperature of the semiconductors. On increasing voltage, first the current remains constant and then the value of current increases suddenly.

PRECAUTIONS

1. In forward bias the *p*-terminal of diode should be connected to positive terminal of the battery and *n*-terminal of diode should be connected to negative terminal of the battery. While in reverse bias the *p*-terminal of diode should be connected to negative terminal of the battery and *n*-terminal of diode should be connected to positive terminal of the battery.
2. All connection should be made carefully.
3. Current should not be passed through the diode for long time it may damage diode.
4. The voltage applied on diode should not be exceeded above its safe limit.

CONCLUSION

VIVAVOCE

1. What is a semiconductor ?
2. What do you mean by majority and minority charge carriers ?
3. What do you mean by *p*-and *n*-type semiconductors ?
4. What are the majority and minority charge carriers in *p*-and *n*-type semiconductor ?
5. What do you mean by forward and reverse biasing ?
6. Explain acceptor and donor ionized atoms.
7. What is potential barrier ? How it is affected by forward and reverse bias ?
8. What do you mean by breakdown condition ?
9. In reverse bias, why current is saturated and so small ?
10. Why breakdown occurs in reverse bias ?
11. What is conductivity ?
12. How semiconductor is different from conductor and insulator ?
13. What is Fermi energy ?
14. Explain energy level diagram in a semiconductor and *p-n* diode.
15. What is depletion region ?
16. What is the direction of electron flow and flow of holes in a diode ?

❑❑❑

EXPERIMENT—29

AIM

To determine the forbidden energy gap of material of P-N junction diode.

APPARATUS REQUIRED

Forbidden energy gap kit, Hot plate, Oil bath and Experimental diode.

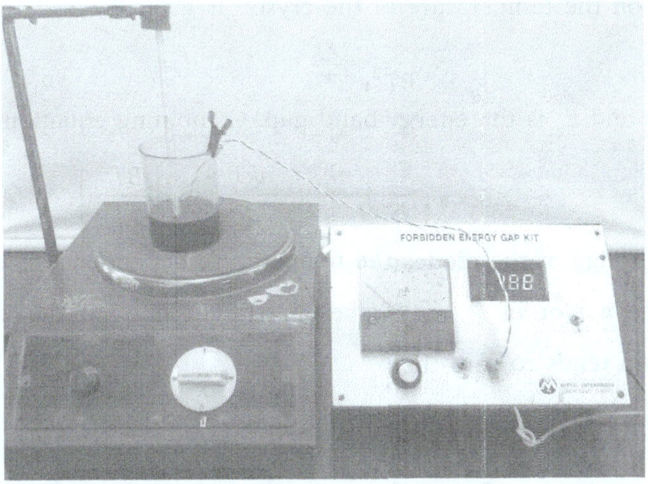

Fig. 29.1. *Forbidden energy gap kit.*

DESCRIPTION OF APPARATUS

The forbidden energy gap kit is basically consisting of a regulated power supply, a digital voltmeter and an analog microampere meter. The regulated variable power supply is used to provide variable potential across the diode in reverse bias. The applied potential can be measured by the digital voltmeter and the analog ammeter measures the current flowing through the diode. The hot plate is used to heat the oil of the oil bath. Oil bath is used to heat the diode to the required temperature. It also has the advantage of being a non-conducting liquid as well as having higher boiling point as compared to water. There are two jacks (J_B, J_D) on the front panel. J_B can be connected to 9.0 V batteries using a plug if the inbuilt power supply fails. The diode under investigation is to be connected to the J_D.

PRINCIPLE

The energy bands in a solid correspond to the energy levels in an atom. An electron in a solid can have only energies that fall within these energy bands. The various energy bands in a solid may or may not overlap depending upon the structure of the solid. If they do not overlap then the intervals between them represent energies which the electrons in the solid cannot have. These intervals are called forbidden bands or energy gaps. If however the adjacent energy bands in a solid overlap, the electrons have a continuous distribution of allowed energies.

THEORY

The current-voltage characteristic of a P-N junction diode (as given in experiment 28) is given by :

$$I = I_S\left(e^{qV/nkT} - 1\right) \quad ...(1)$$

where I is the forward junction current, I_S is reverse saturation current, V is the junction voltage, q is the electronic charge and k is the Boltzmann constant. T is the temperature in Kelvin and n is a constant.

The inverse saturation current depends only on the concentration of minority charge carriers. These carriers are generated in the crystal through intrinsic process. Their rate of generation depends only on the temperature of the crystal is given by

$$I_S = BT^3 e^{-\frac{E_g}{nkT}} \quad ...(2)$$

where B is a constant and E_g is the energy band gap. Combining equations (1) and (2), we get

$$V = \frac{E_g}{q} + \frac{nkT}{q} - \frac{nkT}{q} \cdot \log\left(\frac{BT^3}{I}\right)$$

In the operating range of the diodes the temperature dependence is mainly determined by the term $\frac{nkT}{q}$. Hence a plot of V vs. T gives a straight line. In the above equation if the temperature T in Kelvin tends to zero then

$$V = \frac{E_g}{q}$$

$$\Rightarrow \quad E_g = qV$$

So if the straight line extrapolated to 0°K gives the energy band gap in electron volts.

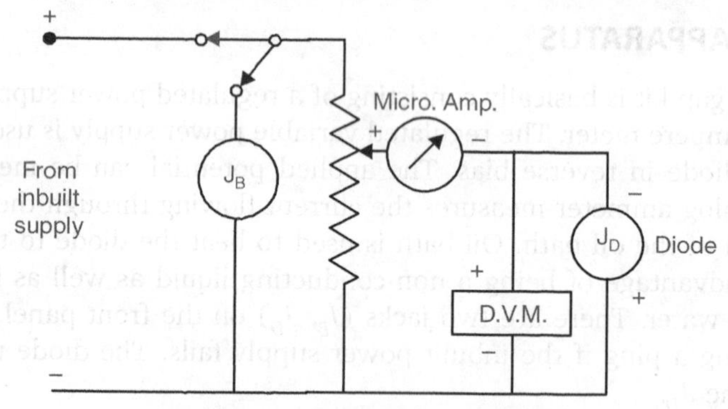

Fig. 29.2. *Experimental arrangement.*

PROCEDURE

1. Switch on the mains supply. You will observe some numbers on the digital panel meter.
2. Connect a 9 V battery to jack J_B, if inbuilt supply fails.
3. Connect the diode under investigation to jack J_D.
4. Adjust the current through the diode for a convenient value; say 80 μA by varying the potentiometer P.

5. Place the diode in an oil bath (The oil bath uses transformer oil or double refined groundnut oil). It is convenient to take oil heated to 120°C (the theory holds good with temperature range 0° to 120°C) in a closed vessel and to place the diode in the oil.
6. Take the voltage across the diode as a function of temp. At each temperature adjust the diode current to the chosen value by adjusting the potentiometer.
7. Allow oil to cool and take at least 10 measurements in the range 25°–120°C.
8. Remove the diode from the oil bath and place it in ice and take a reading.
9. Tabulate the readings in Table 29.1.

OBSERVATION

Fixed Current = ... μA

Table 29.1

Temperature 'T' (in °C)	*Temperature 'T' (in K)*	*Junction voltage in volts*

GRAPH

1. Draw a graph between the junction voltage and absolute temperature as shown below :

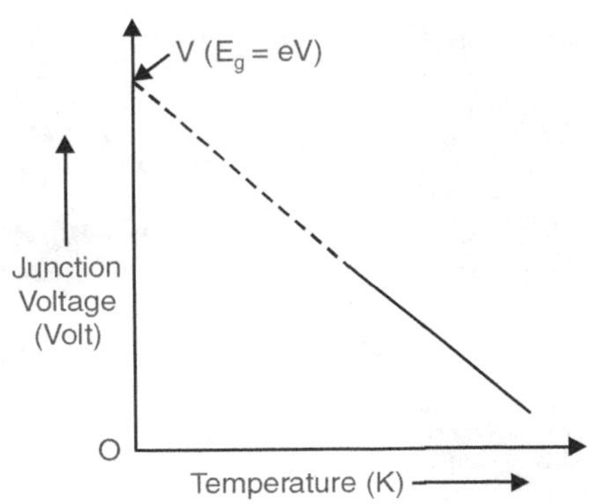

Fig. 29.3. *Graph.*

2. Extrapolate the graph to 0 K. The value expressed in electron volts gives the energy band gap of the material of the diode.

RESULT

When extrapolated the line up to 0 K the line intersects the *Y*-axes at .... Volts. So the forbidden energy gap of the experimental diode is .... eV.

PRECAUTIONS

1. Don't interchange the battery and diode cords.
2. Check the voltage of the battery. It should be 9 V on load.
3. Avoid use of water for temperature variation as any amount of conduction may lead to short-circuiting of the diode.
4. If there is any trouble, please check the polarity of the diode. The positive end of the diode must go to the negative end of the battery.

CONCLUSION

ALTERNATIVE METHOD

APPARATUS REQUIRED

A DC power supply (0 to 15 V), a heating arrangement to heat the diode, a thermometer (0°C to 100°C), a micro-ammeter (0 to 50 μA), and a *p-n* junction diode.

Fig. 29.4. *Forbidden energy gap kit.*

DESCRIPTION OF APPARATUS

The forbidden energy gap kit is basically consisting of a regulated D.C. power supply, a voltmeter and micro-ammeter. The regulated variable power supply is used to provide variable potential across the diode in reverse bias. The applied potential can be measured by the voltmeter and the current flowing through the reverse biased diode is measured with a micro-ammeter. A heating system (oil bath or heating coil) is used to heat the diode to the required temperature.

THEORY

The forbidden energy gap is an energy range in a solid where no electron states can exist. In graphs of the electronic band structure of solids, the band gap generally refers to the energy difference (in electron volts) between the top of the valence band and the bottom of the conduction band in insulators and semiconductors. This is equivalent to the energy required to free an outer shell electron from its orbit about the nucleus to become a mobile charge carrier, able to move freely within the solid material. So the band gap is a major factor determining the electrical conductivity of a solid. When diode is used in reverse biased mode, only minority charge carriers are allowed to flow through the junction and the minority charge carriers are generated through the breaking of the covalent bonds which requires energy equal to band gap energy. So, minority charge carrier concentration varies with temperature, which changes the reverse biased current in the diode.

In the given experiment, variation of minority carriers is carried out through a study of temperature variation of reverse saturation current which is the diode under reverse bias condition.

Reverse saturation current is given by,

$$I_0 = Ae\left[\frac{D_p}{L_p}p_n + \frac{D_n}{L_n}n_p\right] \quad ...(1)$$

where A is the area of the junction,

D_p and D_n are the diffusion coefficients of holes and electrons respectively,

L_p and L_n are the diffusion lengths of holes and electron respectively,

p_n is the concentration of minority holes in *n*-type region, and

n_p is the concentration of minority electrons in *p*-type region.

Using the law of mass action $n_n p_n = n_i^2$ and $p_p n_p = n_i^2$ and $n_n = N_D$ and $p_p = N_A$ into equation (1), we get

$$I_0 = Ae\left[\frac{D_p}{L_p N_D} + \frac{D_n}{L_n N_A}\right]n_i^2 \quad ...(2)$$

The saturation current is thus proportional to the area of the junction A and n_i^2.

But, from law of mass action $= n_i^2 = N_C N_V c^{-E_g/kT}$

$$I_0 = Ae\left[\frac{D_p}{L_p N_D} + \frac{D_n}{L_n N_A}\right]N_C N_V e^{-E_g/kT}$$

$$= Ce^{-E_g/kT} \quad ...(3)$$

where $C = Ae\left[\frac{D_p}{L_p N_D} + \frac{D_n}{L_n N_A}\right]N_C N_V$ is a constant.

It can be seen from equation (3) that the saturation current depends on (E_g/kT) exponentially.

Taking log on both the sides of equation (3), we get

$$\ln I_0 = \ln C - E_g/kT$$

$$2.3026 \log I_0 = 2.3026 \log C - E_g/kT$$

$$\log I_0 = \log C - \frac{E_g}{2.3026 \times kT} \quad \text{...(4)}$$

$$\log I_0 = \log C - \frac{E_g}{2.3026 \times k \times 10^3} \times \frac{10^3}{T} \quad \text{...(5)}$$

From equation (4), we can see that reverse saturation current depends only on the temperature. A plot of $\log I_0$ versus $\frac{10^3}{T}$ gives a negative slope having value of $\frac{E_g}{2.3026 \times k \times 10^3}$.

So, with the help of slope of straight line, band gap E_g can be calculated using,

$$E_g = 2.3026 \times k \times 10^3 \times \text{slope of graph} \left[\frac{PR}{QR}\right] \text{eV} = 0.198 \times \text{slope of graph} \left[\frac{PR}{QR}\right] \text{eV.}$$

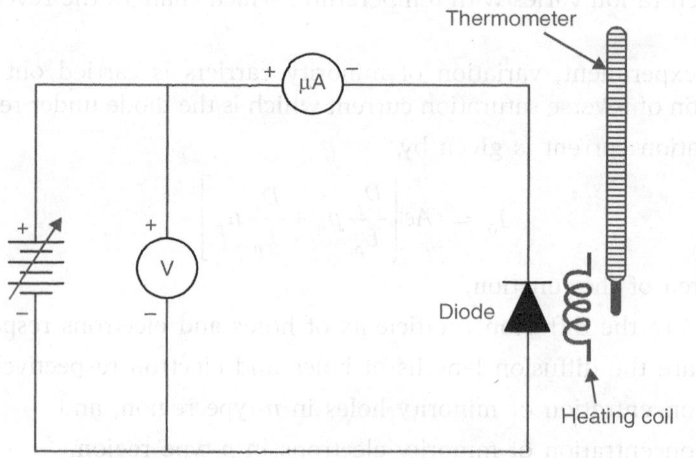

Fig. 29.5. *Circuit diagram.*

PROCEDURE

1. Make the connections as given in the circuit diagram (diode must be connected in reverse bias mode).
2. Now switch on the power supply and voltage should be adjusted to any minimum value say 3 volts. The current through the diode and the room temperature are noted.
3. The heating coil and diode is heated up to 100°C and the corresponding current is noted using micro-ammeter.
4. Now switch-off the power of heating coil.
5. The heating coil is allowed to cool slowly. As its temperature falls, the current through the diode decreases. As the temperature falls through 5°C (or any convenient interval) the corresponding currents should be noted in table.

6. Repeat the steps 1–5 for different reverse bias voltages.
7. Plot graph by taking $1000/T$ on x-axis and log I_0 on y-axis. A straight line such as Fig. 29.6 is obtained.
8. The slope of the straight line is determined and the band gap E_g is calculated using the given formula.

OBSERVATION

Table 29.2

Room Temperature: .... **Reverse Bias Voltage: ....**

S.No.	Current I_0 (μA)	Temperature t (°C)	Temperature $T = t + 273$ (K)	$1000/T$ (K^{-1})	log I_0

CALCULATION

Forbidden energy gap E_g can be calculated using the formula:

$$E_g = 0.198 \times \text{Slope of graph} \left[\frac{PR}{QR}\right] \text{ eV}$$

GRAPH

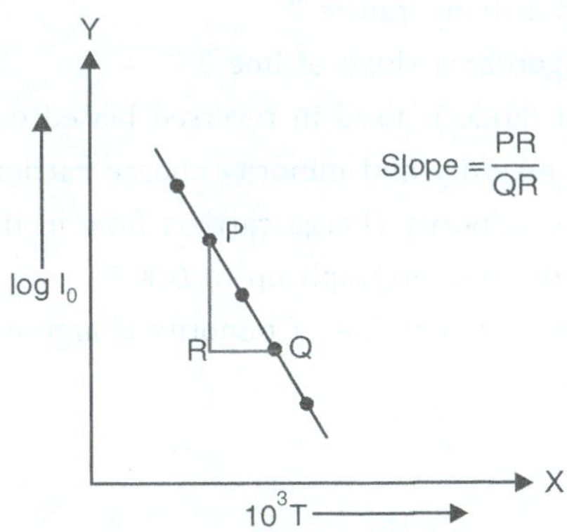

Fig. 29.6

RESULTS

The forbidden Gap of the given *p-n* junction diode is found to be .... eV.

PRECAUTIONS

1. The diode and thermometer should be placed at the heating coil.
2. Diode must be connected in reverse bias mode.

CONCLUSION

VIVAVOCE

1. What do you mean by band gap ?
2. How do you differentiate a good conductor, a semiconductor and an insulator on this band gap ?
3. What is a Fermi level ?
4. How does I_s depend upon temperature ?
5. How do you get band gap as a slope of line ?
6. Why in this experiment diode is used in reversed biased condition ?
7. What do you mean by majority and minority charge carriers ?
8. Tell whether majority or minority charge carriers flow in this experiment and why ?
9. Why do we extraplate the line in graph up to 0 K ?
10. What is the main cause of generation of minority charge carriers ?

❑❑❑

EXPERIMENT—30

AIM

To study the characteristics of a photodiode.

APPARATUS REQUIRED

A photodiode, Voltmeter, Micro/milliammeter, D.C. power supply with Potentiometer (Variable resistance), Lamp with intensity control.

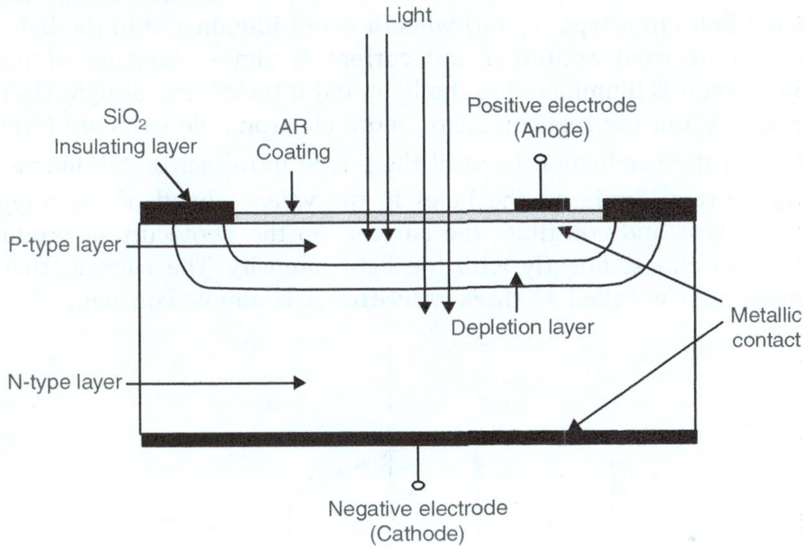

Fig. 30.1. *Structure of photodiode.*

DESCRIPTION

Photodiode

A photodiode consists of a reverse biased *p-n* junction diode embedded in a glass or clear plastic. The upper surface of the photodiode is open to radiation, to absorb maximum radiation falling on the junction as shown in Fig. 30.1. Remaining sides are painted black or enclosed in a metallic case. The whole unit is extremely small and in the order of 2–3 mm size.

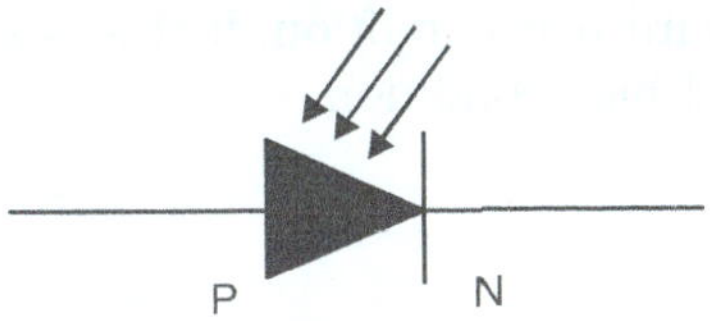

Fig. 30.2. *Circuit symbol.*

PRINCIPLE

A photodiode is a *p-n* junction semiconductor device, which is operated in reverse bias. With no illumination, there exists only the small reverse saturation current (I_0). On illuminating the reverse biased *p-n* junction, new electron-hole pairs are formed, the concentration of which is proportional to the incident light flux. Under large reverse bias condition, the total reverse current I is given by $I = I_S + I_0$, where I_S is the short circuit current and is proportional to the light intensity.

THEORY

The volt-ampere characteristic curve for a typical germanium photodiode is shown in Fig. 30.5. It is different from simple *p-n* junction diode. In photodiode a small current (reverse saturation current) flows same as normal *p-n* junction due to the flow of minority charge carriers (electrons in *p*-type region and holes in *n*-type region) when it is not illuminated to the light. These carriers are generated due to thermal excitation and current is almost constant of the order of few microampere. But when it is illuminated to the light and if the energy of light ($h\nu$) is greater than the band gap energy (E_g) of the semiconductor, more electron-hole pairs are formed. The photo-generated electrons in the conduction band of the *p*-type move across the junction to the *n*-type and constitute the current. Similarly, the holes in the valence band of the *n*-type move across the junction to the *p*-type and constitute the current. So the photocurrent produced due to the radiation (light) varies almost linearly with the light intensity. The current, that flows through the junction without light is called as **dark current,** it is almost constant.

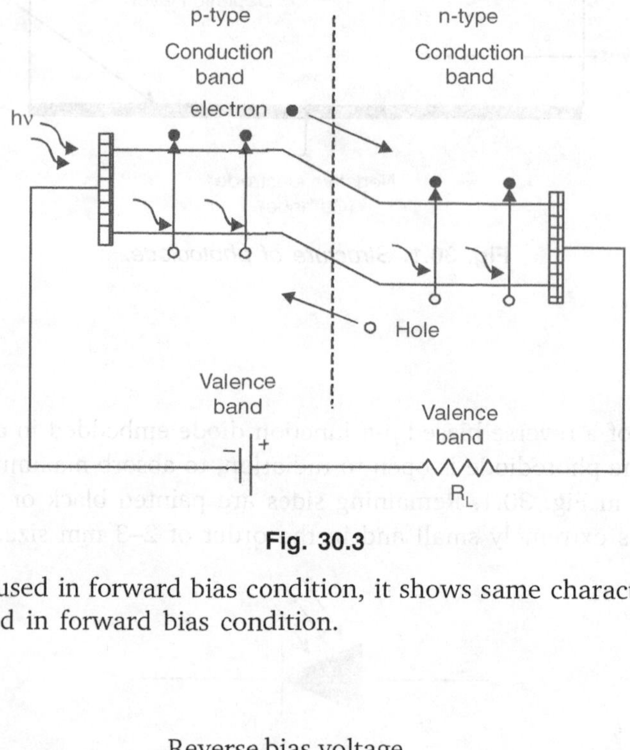

Fig. 30.3

If photodiode is used in forward bias condition, it shows same characteristics as an ordinary diode, so it is not used in forward bias condition.

FORMULA

$$\text{Photodiode dark resistance } (R_0) = \frac{\text{Reverse bias voltage}}{\text{Dark current}}$$

CIRCUIT DIAGRAM

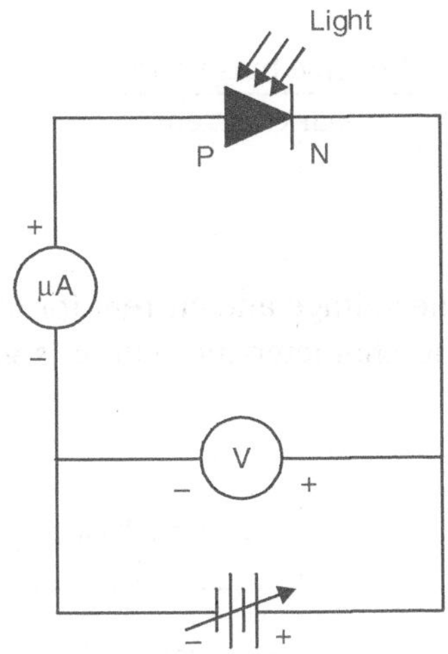

Fig. 30.4. *Circuit diagram.*

PROCEDURE

1. First assemble the circuit as shown in Fig. 30.4.
2. Now switch on the power supply by keeping voltage at minimum.
3. First without illuminating lamp, vary the reverse bias voltage from 0 to 12 volts in regular steps and for each voltage record the value of current. Record all observations in the table.
4. Now switch on the lamp and vary the voltage in steps again. Note the current for each setting and record in the Table 30.1.
5. Now vary the intensity of the lamp and for each intensity repeat the step 4.
6. Now plot a graph (voltage versus current) for every intensity level.
7. Calculate the dark resistance at any bias voltage using the above given formula.

OBSERVATIONS

Table 30.1

S.No.	*Reverse Bias Voltage (Volts)*	*Intensity I_0 (Zero Intensity)*	*Intensity I_1*	*Intensity I_2*	*Intensity I_3*	*Intensity I_4*
		Current	*Current*	*Current*	*Current*	*Current*

CALCULATION

$$\text{Photodiode Dark Resistance } (R_0) = \frac{\text{Reverse bias voltage}}{\text{Dark current}} = \text{..............} \text{ ohm}$$

GRAPH

The graph can be plotted between the voltage and current for different intensity level. Take voltage on *X*-axis and current on *Y*-axis. The characteristics curve is given in Fig. 30.5.

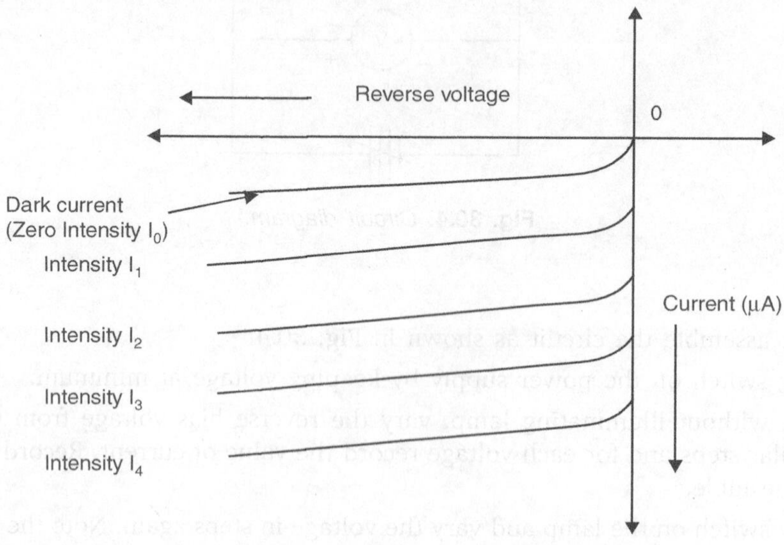

Fig. 30.5. *Characteristics of a photodiode.*

RESULTS

From graph it is clear that the reverse current of a photodiode depends on the intensity of the incident light. Current varies linearly with the light intensity. On increasing the intensity of light the current also increases, which shows that the light produces extra electron-hole pairs on the photodiode. Due to the flow of these electrons and holes (minority charge carriers) the reverse current increases.

PRECAUTIONS

1. It should be noted that photodiode always works in reverse bias condition, so check the biasing of the diode before the power is switch on.
2. All the readings should be noted carefully.

CONCLUSION

VIVAVOCE

1. What is a photodiode ?
2. Why it is called photodiode ?
3. Why it is advantageous over ordinary bulk type photoconductive cell ?
4. Why only minority charge carriers are taken into account for short circuit current I_S ?
5. How the current varies with radiation flux ?
6. What are the applications of photodiode ?
7. What is reverse saturation current ?
8. Explain the working of photodiode.
9. Give the name of material, which is used for making photodiode.
10. State whether photodiode is used in forward bias or reverse bias mode and why ?

❑❑❑

EXPERIMENT—31

AIM

To study the characteristic curve of solar cell.

APPARATUS REQUIRED

Solar cell panel, voltmeter, milliammeter, variable resistance. Lamp with intensity control.

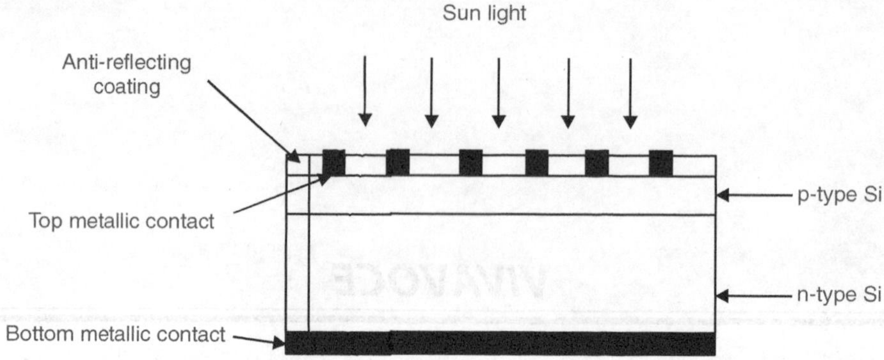

Fig. 31.1. *Structure of solar cell.*

DESCRIPTION

Solar Cell

A solar cell is semiconductor device (*p-n* junction), which converts radiant energy of sunlight directly and efficiently into electrical energy. The conversion of optical energy into electrical energy is called as photovoltaic effect, so a solar cell is also called as photovoltaic cell.

Solar cell consists of a shallow *p-n* junction formed below the surface of a silicon crystal. The energy gap in silicon is ≈ 1.12 eV. There is an ohmic contact at the top in the form of a comb like structure which covers above 7% of silicon area, and an ohmic contact at the bottom which covers the entire bottom area. To absorb more sunlight, an antireflecting coating is made over the top surface. This causes generation of an electrical potential difference across the top and the bottom of the cell, which deliver power to an external circuit.

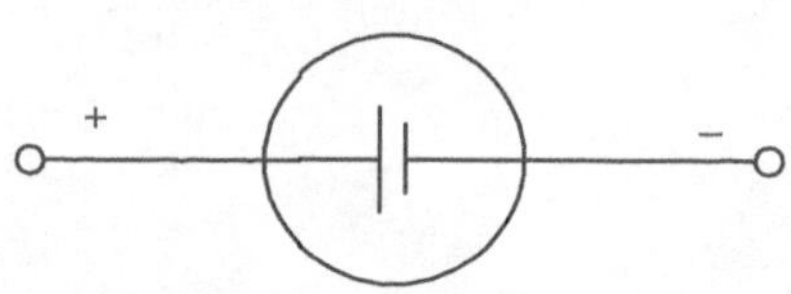

Fig. 31.2

PRINCIPLE

It is possible to grow a single crystal containing both *p*-type and *n*-type regions. The transition zone between them is called a *p-n* junction. When light is incident on this crystal, electron-hole pairs are produced in the crystal and the electron drift into the n-region while the holes drift to the p-region. In the absence of an externally applied field, the light induced current develops a photo-voltage across the crystal, hence the phenomena is called the photovoltaic effect. Such a crystal obviously can be used as a battery whose energy is derived from, say visible light.

THEORY

A beam of photons from a lamp is incident on the depletion region of an unbiased p-n junction, creates large number of electrons and holes pair in the cell. Due to diffusion of the charge carriers, the charge carriers tend to move towards the junction but due to the presence of internal electric field of potential barrier the majority charge carriers are repelled back. Internal electric field does not allow the flow of electrons (majority charge carriers) from n to p region and the flow of holes (majority charge carriers) from p to n region. But internal electrical field forces the electrons (minority charge carriers) from p to n region and holes (minority charge carriers) from n to p-region. The two sides of the junction become oppositely charged due to the accumulation of the electrons on the n-region and holes on the p-side. This produces a potential difference, called as photo emf. When any external circuit is connected across the solar cell, this voltage drives a current around the circuit. Thus the solar cell acts as a cell or battery. The n side behaves like negative terminal and p-side behaves like positive terminal.

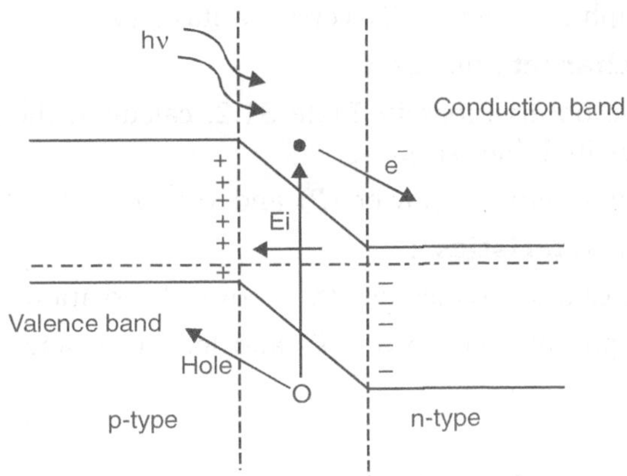

Fig. 31.3

CIRCUIT DIAGRAM

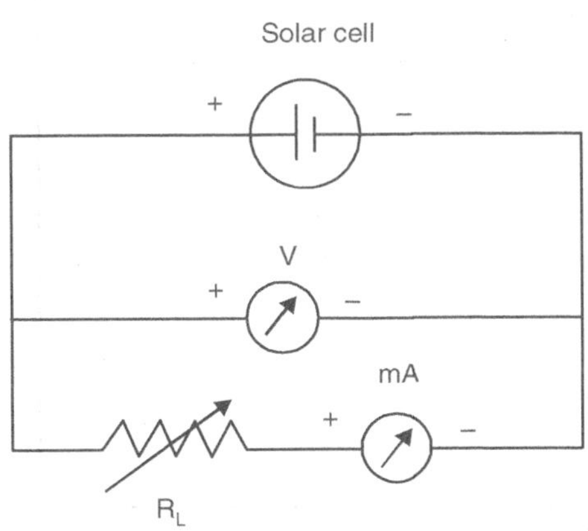

Fig. 31.4. *Circuit diagram.*

PROCEDURE

A. Illumination characteristics :

1. Connect the circuit as shown in Fig. 31.4.
2. Now turn on the lamp and keep the intensity control at its minimum.
3. Now keep the load R_L at 0 ohm and vary the intensity of lamp in steps.
4. Record the current in milliammeter each time and note it down in the Table 31.1.
5. Plot a graph between the current (I) and intensity of lamp (voltage across the lamp).

B. Current–voltage (I-V) characteristics :

1. First keep the intensity of the lamp at the minimum.
2. Now set the value of load resistance at 0 ohm and record the value of voltage in voltmeter and current in milliammeter in Table 31.2.
3. Now increase the value of load resistance in steps for the same intensity of the lamp. Again record the corresponding value of voltage and current in Table 31.2.
4. Repeat above procedure for different intensity of the lamp.
5. Now set the intensity to the maximum and repeat the step 3.
6. Now plot a graph for current (I) versus voltage (V).

C. Power-voltage characteristics :

1. Using the above observations in Table 31.2, calculate the output power $P = VI$ and enter the values in Table 31.3.
2. Now plot a graph between power (P) and voltage (V).

D. Power-load characteristics :

1. For Power-load Characteristics use the above observations in Table 31.3.
2. Now plot a graph between power (P) and load resistance (R_L).

OBSERVATION

Table 31.1. Illumination characteristics.

S.No.	***Intensity (Volts)***	***Current (mA)***

Table 31.2. I-V Characteristics.

S.No.	R_L (Ohm)	*Intensity 1 (Zero intensity)*		*Intensity 2*		*Intensity 3*		*Intensity 4 (Full intensity)*	
		Voltage (V)	*Current (mA)*	*Voltage (V)*	*Current (mA)*	*Voltage (V)*	*Current (mA)*	*Voltage (V)*	*Current (mA)*

Table 31.3. Power-voltage and power-load characteristics.

S.No.	R_L (Ohm)	*Intensity 1 (Zero intensity)*		*Intensity 2*		*Intensity 3*		*Intensity 4 (Full intensity)*	
		Voltage (V)	*Power (mW)*	*Voltage (V)*	*Power (mW)*	*Voltage (V)*	*Power (mW)*	*Voltage (V)*	*Power (mW)*

GRAPH

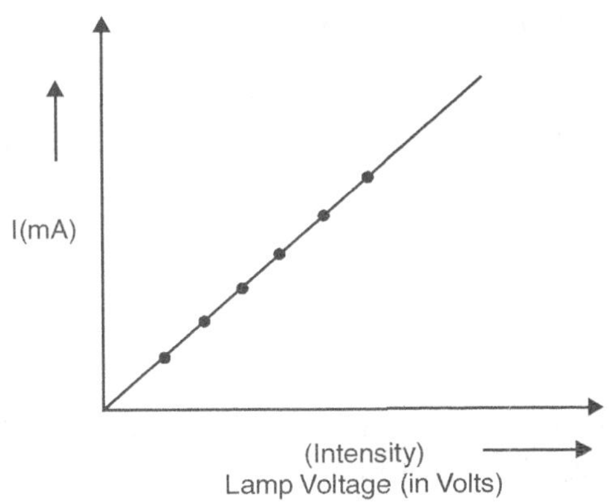

Fig. 31.5. *Illumination characteristics.*

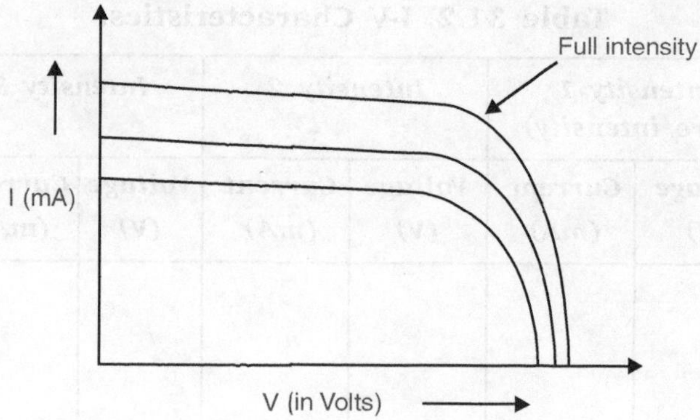

Fig. 31.6. *I-V characteristics.*

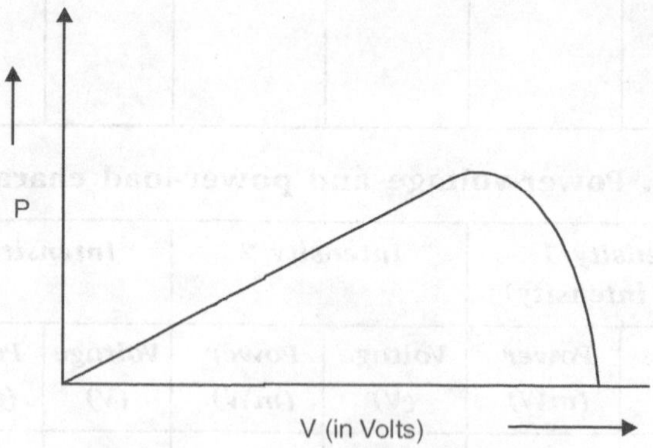

Fig. 31.7. *Power-voltage characteristics.*

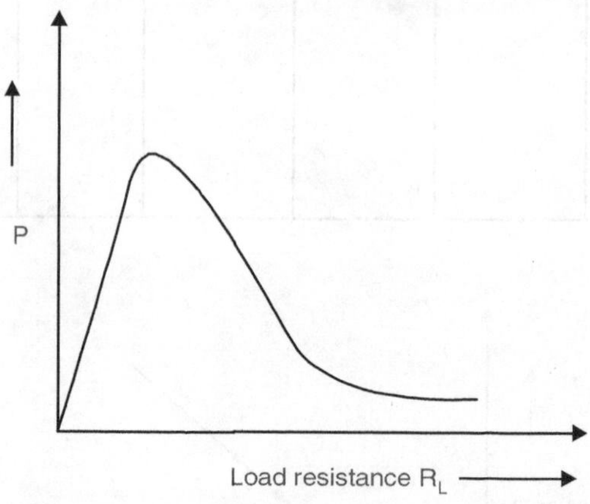

Fig. 31.8. *Power-load characteristics.*

RESULT

The graph is plotted between various parameters for the given solar cell. Fig. 31.5 represents Illumination Characteristics, Fig. 31.6 represents I-V Characteristics, Fig. 31.7 represents Power-Voltage characteristics and Fig. 31.8 represents Power-Load characteristics of the given solar cell.

PRECAUTIONS

1. All connection should be made correctly. Positive terminal should be connected to positive and negative-to-negative.
2. All readings should be taken carefully.

CONCLUSION

VIVAVOCE

1. What do you mean by photovoltaic cell ?
2. Why is it called solar cell ?
3. Explain the working of solar cell.
4. How the solar cell is fabricated ? Explain.
5. What is the use of solar cell ?
6. For making of solar cell which material is used ?
7. Why antireflection coating is made over the top surface ?
8. What do you mean by open circuit voltage of the solar cell ?
9. Why electrons get accumulated on the n-region and holes on the p-region ?
10. What do you mean by diffusion of charge carriers ?
11. How is internal electric field generated ?
12. Show the circuit symbol of a solar cell.

❑❑❑

EXPERIMENT—32

AIM

To study the characteristics of the given NPN (or PNP) transistor in the common emitter (C.E.) mode.

APPARATUS REQUIRED

A NPN (or PNP) junction transistor, Two regulated power supply with 0 to 12 volt output voltage, One milliammeter, One microammeter, Two voltmeters, Two rheostat or potentiometer, Connecting wires.

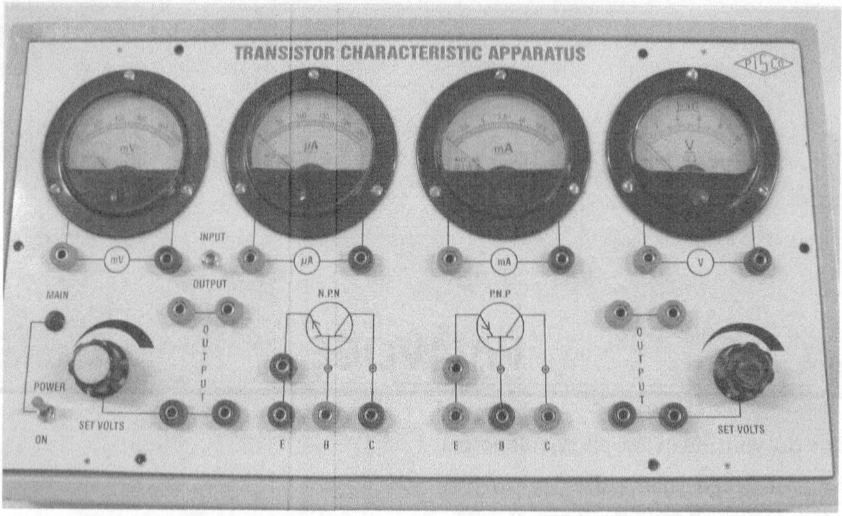

Fig. 32.1. *Transistor characteristic apparatus.*

DESCRIPTION OF THE APPARATUS

A NPN (or PNP) transistor in common emitter (C.E.) mode is shown in Fig. 32.4. Two power supplies used here for biasing input and output regions. With the help of one rheostat (or variable resistance) we can vary the forward bias voltage V_{BE} and reverse bias voltage V_{CE} is varied with the help of other rheostat (or variable resistance). These voltages are read by the two voltmeters. Similarly I_B is read by the microammeter connected in series and I_C is read by the milliammeter connected in series.

The arrows show the direction of conventional current (positive charge or hole movement) between the base and emitter, so electrons would flow in the opposite direction. In an actual transistor, the collector terminal is displaced more than the other for recognition or has a dot near it. The base is forward biased by connecting to the positive terminal of a variable battery whose negative terminal is connected to the emitter. The collector is reverse biased by connecting it to the positive terminal of another variable battery whose negative terminal is connected to the emitter.

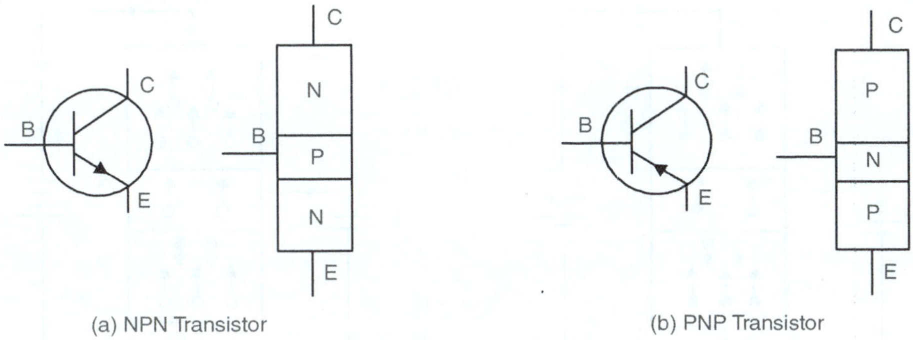

Fig. 32.2. *Symbolic diagram.*

PRINCIPLE

A junction transistor is a semiconductor device consisting of two *p-n* junctions placed back to back, one under a forward bias and other under a reverse bias. A forward biased *p-n* junction offers a low resistance, while a reverse biased junction offers a high resistance. Therefore, a small signal applied across the forward biased junction would appear with a much higher voltage across the reverse biased junction. The device can thus act as an amplifier. It is a bipolar device because in it both majority and minority carriers play significant roles. The curves representing the variation of current with voltage in a transistor circuit are called transistor characteristic.

THEORY

A transistor is a three terminal device having three regions. The central region is called the base and outer regions are called emitter and collector. Emitter and collector are made by the same type of semiconductor (*p*-type or *n*-type) and the base is different from outer regions. If the base is *p*-type and collector-emitter are *n*-type, then it is called as NPN transistor. Similarly if the base is *n*-type and collector-emitter are *p*-type, then it is called as PNP transistor. Emitter is highly doped because it provides large number of charge carriers. Base is made thin and lightly doped because it has to pass most of the charge carriers coming from the emitter to collector. Collector is made larger in width and moderately doped because it collects charge carriers coming from the emitter.

Transistor can be Operated in Three Modes

1. Common Base mode (Base is made common to both the input and output).
2. Common Emitter mode (Emitter is made common to both the input and output).
3. Common Collector mode (Collector is made common to both the input and output).

In common emitter operation, the base-emitter junction is forward biased and the collector-emitter junction is reverse biased. In an NPN transistor, when a forward bias voltage is applied to the base-emitter junction (input junction). It allows donor electrons (majority charge carriers) to inject into the base region. It creates large concentration difference in the base region. These electrons diffuse through the base from the high concentration region near the emitter towards the low concentration region near the collector. So a current I_C flows in the collector circuit. The electrons in the base are called minority charge carriers because the base is made of *p*-type semi-conductor. The base region of the transistor is made always thin, so that carriers can easily diffuse across it in much less time than the semiconductor's minority carrier lifetime. It minimizes the

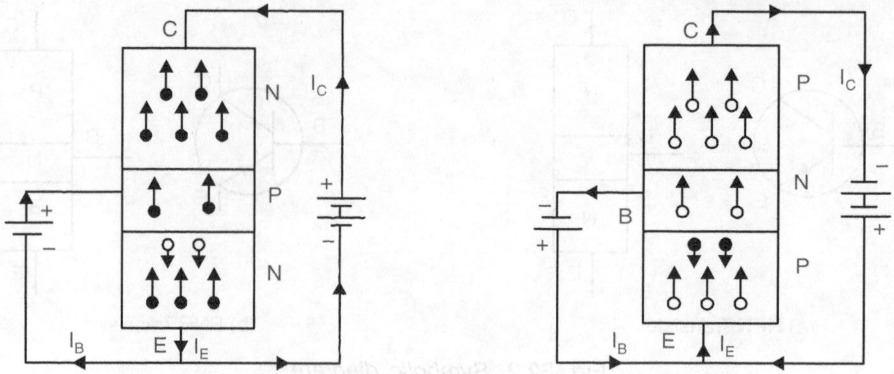

(a) NPN Transistor (b) PNP Transistor

Fig. 32.3. *Current flow diagram.*

percentage of carriers that recombine before reaching the collector-base junction. So a small current I_B is obtained here. The collector-base junction is reverse biased, so electrons (minority carriers in base region) that diffuse through the base towards the collector are swept into the collector by the internal electric field in the depletion region of the collector-base junction. From Kirchhoff's first law, it follows that always, if I_E is the emitter current $I_E = I_C + I_B$. Thus in this way charge carriers are easily transferred from emitter to collector via base region (from lower resistance region to higher resistance region *i.e.,* transfer of resistance).

CHARACTERISTICS OF TRANSISTOR

There are four variables base voltage V_{BE} and base current I_B for the input circuit and collector voltage V_{CE} and collector current I_C for the output circuit. The base circuit is a low current circuit of the order of microampere and the collector circuit is a relatively high current circuit.

1. **Input characteristics.** In the common emitter mode the input characteristics is determined by keeping V_{CE} constant and input base current I_B is noted for each value of V_{BE}. In input characteristics, graph is plotted between base-emitter voltage V_{BE} and base current I_B.
2. **Output characteristics.** Output characteristic graph is plotted between collector emitter voltage V_{CE} and collector current I_C by taking the base current I_B constant. In this manner a family of curves is obtained.
3. **Current transfer characteristics.** Current transfer characteristic graph is plotted between collector current I_C and base current I_B. By keeping V_{CE} constant, I_C is determined for each value of I_B. It is also a straight line.

   In this graph we can determine current gain β by the formula

$$\beta = \left(\frac{\Delta I_C}{\Delta I_B}\right)_{V_{CE}}$$

   Remember β is always greater than unity, because I_C is always greater than I_B.

FORMULA USED

Current gain in common emitter mode is given by

$$\beta = \left(\frac{\Delta I_C}{\Delta I_B}\right)_{V_{CE}}$$

CIRCUIT DIAGRAM

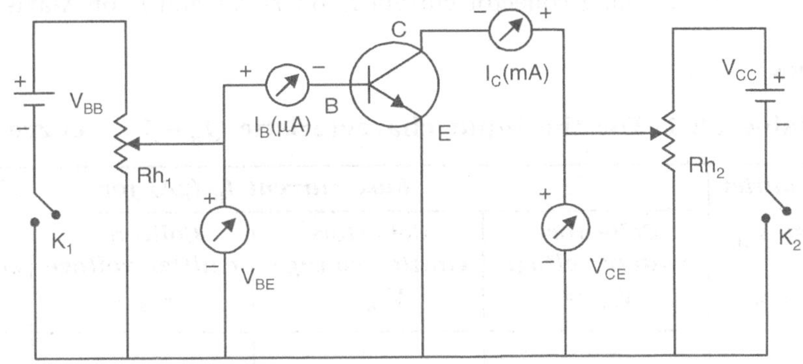

Fig. 32.4. *NPN transistor in common emitter (C.E.) mode.*

PROCEDURE

1. To obtain the input characteristic (I_B – V_{BE}) curve

(*i*) Adjust the rheostat Rh_2 to obtain the collector-emitter voltage V_{CE} = 0 V.

(*ii*) For this value of V_{CE}, set the value of base-emitter voltage V_{BE} to any particular value with the help of rheostat Rh_1 and note the corresponding base current I_B from the micro-ammeter.

(*iii*) Now increase the value of base-emitter voltage V_{BE} in steps and note the corresponding base current I_B for each value of V_{BE}.

(*iv*) Repeat the above steps 1–3 for different constant value of V_{CE}.

(*v*) Now plot a graph by taking the base-emitter voltage V_{BE} on the *X*-axis and the base current I_B on *Y*-axis.

2. To obtain the output characteristic (I_C – V_{CE}) curve

(*i*) Adjust the rheostat Rh_1 to obtain the base current I_B at constant value say 20 μA in the micro-ammeter.

(*ii*) Now for this constant value of I_B adjust the collector voltage V_{CE} with the help of rheostat Rh_2 and measure the corresponding collector current I_C with the help of milli-ammeter.

(*iii*) Now increase the value of collector voltage V_{CE} in steps and measure the corresponding values of collector current I_C.

(*iv*) Repeat the above steps 1–3 for different constant value of base current I_B.

(*v*) Now plot a graph by taking the collector-emitter voltage V_{CE} on *X*-axis and collector current I_C on *Y*-axis.

3. To obtain the current transfer characteristic (I_B – I_C) curve

(i) First adjust the rheostat Rh_2 to obtain the collector-emitter voltage V_{CE} to any fixed value.

(ii) Now set the base current I_B with the help of rheostat Rh_1 at any lower value and note the collector current I_C.

(iii) Now increase the value of base current I_B with the help of rheostat Rh_1 and note the corresponding value of collector current I_C.

(iv) Plot a graph by taking collector current I_C on Y-axis and I_B on X-axis.

OBSERVATIONS

Table 32.1. For the input characteristic (I_B – V_{BE}) curve.

S.No.	Base-emitter voltage V_{BE} (in Volt)	Base current I_B (μA) for			
		Collector-emitter voltage V_{CE} =	Collector-emitter voltage V_{CE} =	Collector-emitter voltage V_{CE} =	Collector-emitter voltage V_{CE} =

Table 32.2. For the output characteristic (I_C – V_{CE}) curve.

S.No.	Collector-emitter voltage V_{CE} (in Volt)	Collector current I_c (mA) for			
		Base Current I_B =	Base Current I_B =	Base Current I_B =	Base Current I_B =

Table 32.3. For the current transfer characteristic (I_B – I_C) curve.

For constant V_{CE} = ...

S.No.	*Base current I_B (μA)*	*Collector current I_C (mA)*

GRAPH

1. Input characteristic graph is plotted between base-emitter voltage V_{BE} and base current I_B by taking the base-emitter voltage V_{BE} on the *X*-axis and the base current I_B on *Y*-axis.
2. Output characteristic graph is plotted between collector-emitter voltage V_{CE} and collector current I_C by taking the collector-emitter voltage V_{CE} on *X*-axis and collector current I_C on *Y*-axis.
3. Current transfer characteristic graph is plotted between collector current I_C and base current I_B by taking collector current I_C on *Y*-axis and base current I_B on *X*-axis.

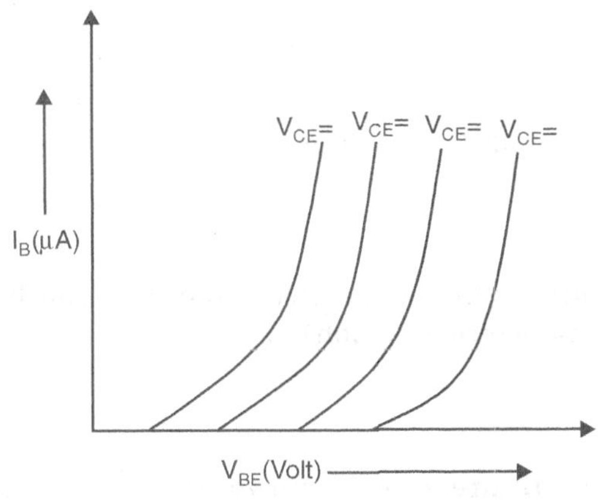

Fig. 32.5. *Input characteristics.*

The curves obtained in this graph are similar to the forward bias characteristics curve of a *p-n* junction diode. This is because in this case the base-emitter region is forward biased. Value of I_B is of the order of microampere, because I_B is the small fraction of I_E. It is seen that the base current I_B increases non-linearly with increase in base-emitter voltage V_{BE}. Also, the input characteristics are slightly dependent upon the collector-emitter voltage V_{CE}. The input resistance is higher compared to that in common base configuration.

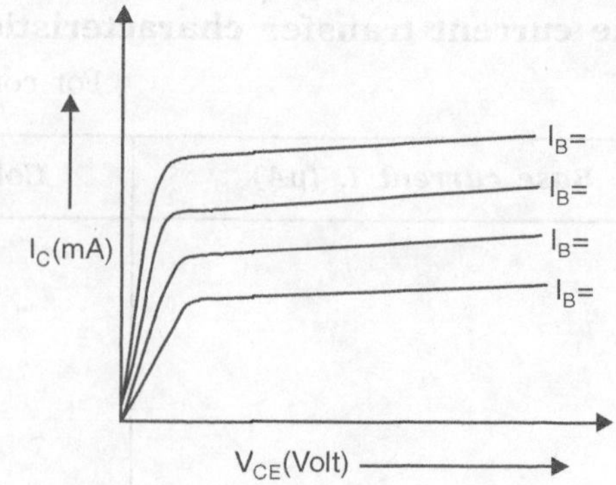

Fig. 32.6. *Output characteristics.*

We can see that collector current I_C increases rapidly by increasing collector-emitter voltage V_{CE}. There is a small collector current even when the base current I_B is zero, this is due to the intrinsic conduction inherent in semiconductors and is highly temperature dependent. But when collector-emitter voltage V_{CE} is decreased below the value of base-emitter voltage V_{BE}, then I_C sharply reduces to zero, because in this case the collector-emitter region becomes forward biased. The forward bias repels further flow of minority charge carriers and we know that I_C is mainly due to the flow of minority charge carriers, so I_C suddenly drops to zero.

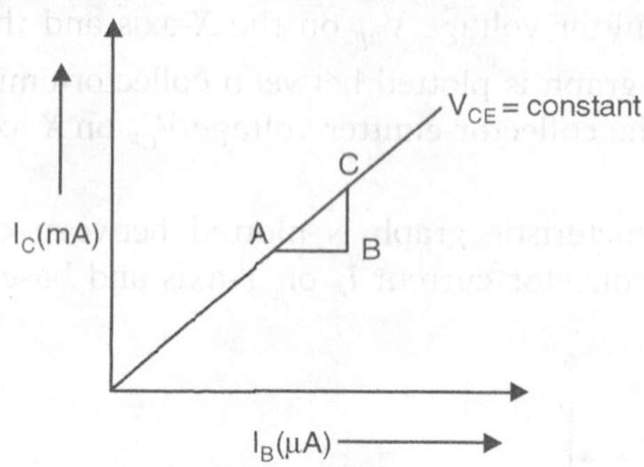

Fig. 32.7. *Current transfer characteristics.*

CALCULATION

Current gain in common emitter mode of the given transistor can be calculated by the slope of the graph (current transfer characteristic graph).

Current Gain :

$$\beta = \left(\frac{\Delta I_C}{\Delta I_B}\right)_{V_{CE}} = \textbf{Slope of the } \mathbf{I_C - I_B} \textbf{ curve} = \frac{CB}{AB} = \ldots\ldots\ldots\ldots$$

RESULT

For the given transistor, in the C.E. configuration. Fig. 32.5 represents the input characteristic curve; Fig. 32.6. represents the output characteristic curve and Fig. 32.7 represents the current transfer characteristic curve.

Current gain in common emitter mode β = ...

PRECAUTIONS

1. Forward bias should be applied between emitter-base junction and reverse bias should be applied between collector-base junction.
2. The transistor should not be used for long time continuously, it may damage due to heating.

CONCLUSION

VIVAVOCE

1. What is a junction transistor ?
2. What do you mean by p and n type semiconductor ?
3. Why is a transistor, so named ?
4. Why it is called bipolar junction ?
5. What are the types of transistor ?
6. Explain the function of base, emitter and collector.
7. Why the base is made thin and is doped less ?
8. Why collector is broad ?
9. Why emitter is heavily doped ?
10. How can check whether a transistor is correct or not ?
11. What do mean by current gain ?
12. Give the relationship between the current gain in C.B. mode and current gain in C.E. mode.
13. Explain how a transistor is biased for proper working ?
14. Explain input and output characteristics of transistor.
15. How transistor acts as an amplifier ?
16. What do you mean by biasing a transistor ?
17. Define static resistance and dynamic resistance.

❑❑❑

EXPERIMENT—33

AIM

To study the characteristics of the given NPN (or PNP) transistor in the common base (C.B.) mode.

APPARATUS REQUIRED

A NPN (or PNP) junction transistor, Two regulated power supply with 0 to 12 volt output voltage, Two milliammeters, Two voltmeters, Two rheostat or potentiometer and connecting wires.

DESCRIPTION

An NPN (or PNP) transistor in common base (C.B.) mode is shown in Fig. 33.1. Two power supplies used here for biasing input and output regions. With the help of one rheostat (or variable resistance) we can vary the forward bias voltage V_{EB} and reverse bias voltage V_{CB} is varied with the help of other rheostat. These voltages are read by the two voltmeters. Similarly I_E and I_C are read by the two milliammeters connected in series.

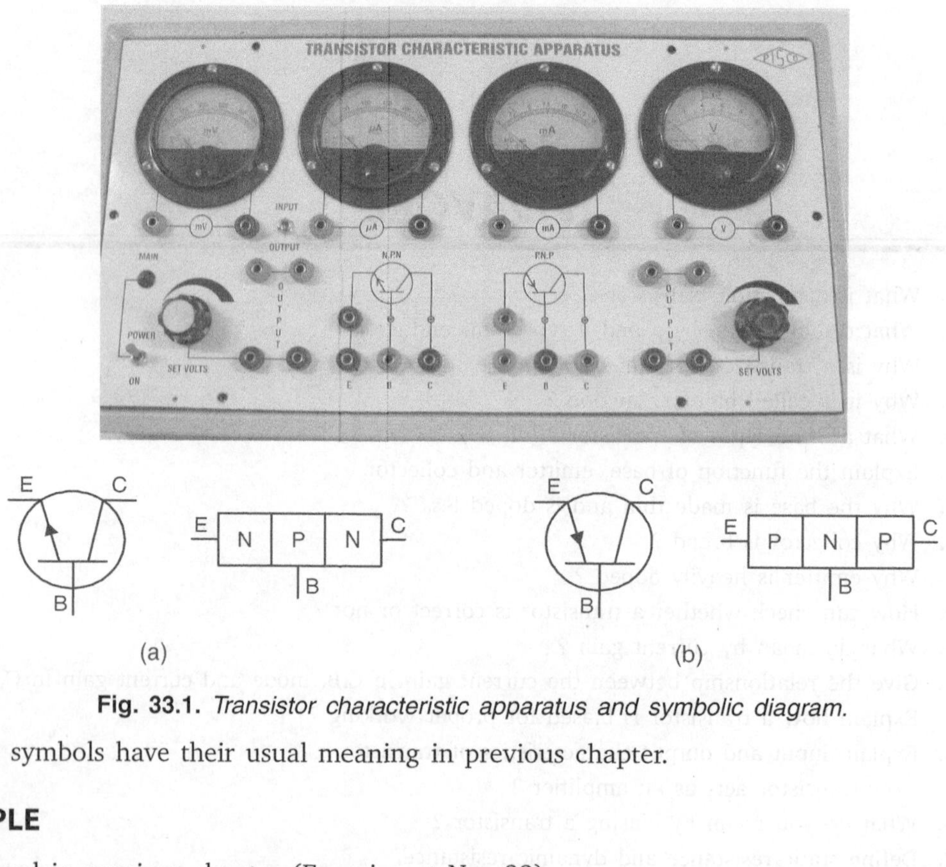

Fig. 33.1. *Transistor characteristic apparatus and symbolic diagram.*

The symbols have their usual meaning in previous chapter.

PRINCIPLE

As discussed in previous chapter (Experiment No.–32).

THEORY

As discussed in previous chapter (Experiment No.–32).

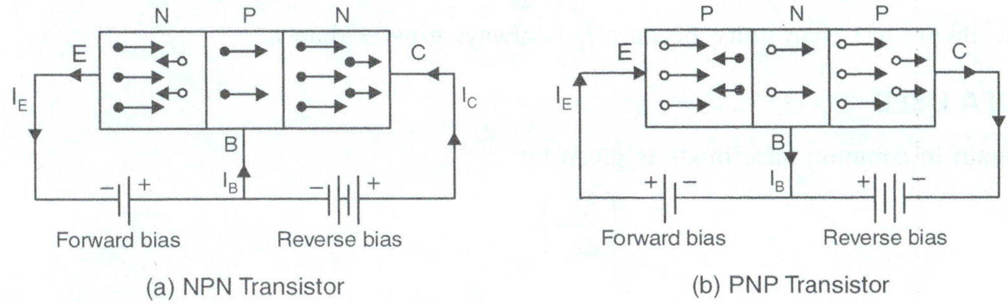

Fig. 33.2. *Current flow diagram.*

In common base general operation, the emitter-base junction is forward biased and the base-collector junction is reverse biased. In an NPN transistor, when a forward bias voltage is applied to the base-emitter junction (input junction), it allows donor electrons (majority charge carriers) to inject into the base region. It creates large concentration difference in the base region. These electrons diffuse through the base from the high concentration region near the emitter towards the low concentration region near the collector. So a current I_C flows in the collector circuit. The electrons in the base are called *minority charge carriers* because the base is made of *p*-type semi-conductor. The base region of the transistor is made always thin, so that carriers can easily diffuse across it in much less time than the semiconductor's minority carrier lifetime. It minimizes the percentage of carriers that recombine before reaching the collector-base junction. So a small current I_B is obtained here. The collector-base junction is reverse biased, so electrons (minority carriers in base region) that diffuse through the base towards the collector are swept into the collector by the internal electric field in the depletion region of the collector-base junction. From Kirchhoff's first law, it follows that always, if I_E is the emitter current $I_E = I_C + I_B$. Thus in this way charge carriers are easily transferred from emitter to collector region via base region (from lower resistance region to higher resistance region *i.e.,* transfer of resistance).

CHARACTERISTICS OF TRANSISTOR

There are four variables base voltage V_{EB} and emitter current I_E for the input circuit and collector voltage V_{CB} and collector current I_C for the output circuit. The base circuit is a low current circuit of the order of microampere and the collector circuit is a relatively high current circuit.

1. **Input characteristics.** In the common base mode the curves showing the variation of the emitter current I_E with emitter voltage V_{EB} at constant V_{CB} are the input characteristics.
2. **Output characteristics.** Output characteristic graph is plotted between collector-base voltage V_{CB} and collector current I_C by taking the emitter current I_E constant. To keep I_E constant V_{EB} is to be varied as V_{CB} is changed. In this manner a family of curves can be obtained.
3. **Current transfer characteristics.** Current transfer characteristic graph is plotted between collector current I_C and emitter current I_E. By keeping V_{CB} constant, I_C is determined for each value of I_E. It is straight line because I_C directly depends on the value of I_E.

In this graph we can determine current gain α by the formula

$$\alpha = \left(\frac{\Delta I_C}{\Delta I_E}\right)_{V_{CB}}$$

α is always less than unity, because I_E is always greater than I_C.

FORMULA USED

Current gain in common base mode is given by

$$\alpha = \left(\frac{\Delta I_C}{\Delta I_E}\right)_{V_{CB}}.$$

CIRCUIT DIAGRAM

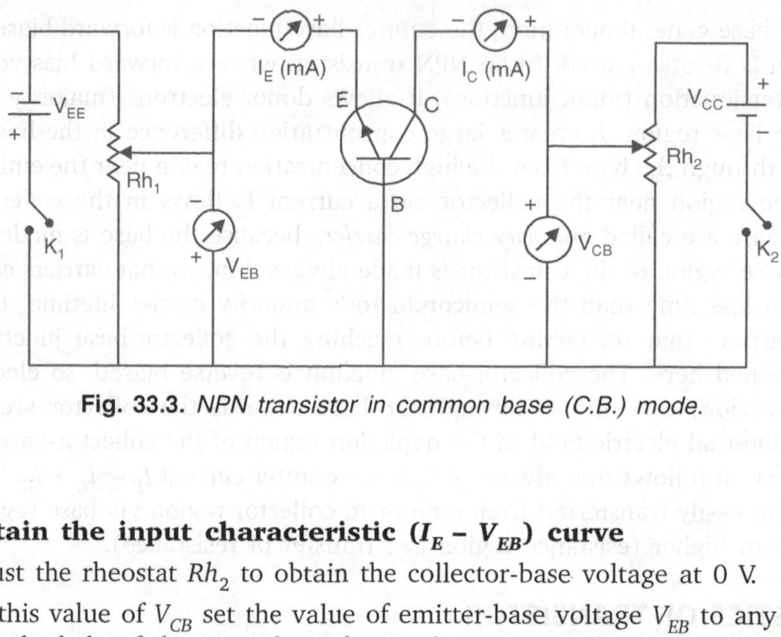

Fig. 33.3. *NPN transistor in common base (C.B.) mode.*

PROCEDURE

1. To obtain the input characteristic (I_E – V_{EB}) curve

(*i*) Adjust the rheostat Rh_2 to obtain the collector-base voltage at 0 V.

(*ii*) For this value of V_{CB} set the value of emitter-base voltage V_{EB} to any particular value with the help of rheostat Rh_1 and note the corresponding emitter current I_E from the milliammeter.

(*iii*) Now increase the value of emitter-base voltage V_{EB} in steps and note the corresponding emitter current I_E for each value of V_{EB}.

(*iv*) Repeat the above steps 1–3 for different constant value of V_{CB}.

(*v*) Now plot a graph by taking the emitter-base voltage V_{EB} on the *X*-axis and the emitter current I_E on *Y*-axis.

2. To obtain the output characteristic (I_C–V_{CB}) curve

(*i*) Adjust the rheostat Rh_1 to obtain the emitter current I_E at constant value say 1 mA in the milliammeter.

(*ii*) Now for this constant value of I_E adjust the collector voltage V_{CB} with the help of rheostat Rh_2 and measure the corresponding collector current I_C with the help of second milliammeter.

(iii) Now increase the value of collector voltage V_{CB} in steps and measure the corresponding values of collector current I_C.

(iv) Repeat the above steps 1–3 for different constant value of emitter current I_E.

(v) Now plot a graph by taking the collector base voltage V_{CB} on X-axis and collector current I_C on Y-axis.

3. To obtain the current transfer characteristic (I_E - I_C) curve

(i) First adjust the rheostat Rh_2 to obtain the collector base voltage V_{CB} equal to any fixed value.

(ii) Now set the emitter current I_E with the help of rheostat Rh_1 at any lower value and note the collector current I_C.

(iii) Now increase the value of emitter current with the help of rheostat Rh_1 and note the corresponding value of collector current I_C.

(iv) Plot a graph by taking collector current I_C on Y-axis and I_E on X-axis.

OBSERVATIONS

Table 33.1. For the input characteristic ($I_E - V_{EB}$) curve.

S.No.	Emitter-base voltage V_{EB} (in Volt)	Emitter current I_E (mA) for			
		Collector-base voltage V_{CB} =	Collector-base voltage V_{CB} =	Collector-base voltage V_{CB} =	Collector-base voltage V_{CB} =

Table 33.2. For the output characteristic ($I_C - V_{CB}$) curve.

S.No.	Collector-base voltage V_{CB} (in Volt)	Collector current I_C (mA) for			
		Emitter current I_E =	Emitter current I_E =	Emitter current I_E =	Emitter current I_E =

Table 33.3. For the current transfer characteristic ($I_E - I_C$) curve.

For constant V_{CB} = ...

S.No.	Emitter current I_E (mA)	Collector current I_C (mA)

GRAPH

1. Input characteristic graph is plotted between emitter-base voltage V_{EB} and emitter current I_E by taking the emitter-base voltage V_{EB} on the X-axis and the emitter current I_E on Y-axis.
2. Output characteristic graph is plotted between collector base voltage V_{CB} and collector current I_C by taking the collector base voltage V_{CB} on X-axis and collector current I_C on Y-axis.
3. Current transfer characteristic graph is plotted between collector current I_C and emitter current I_E by taking collector current I_C on Y-axis and emitter current I_E on X-axis.

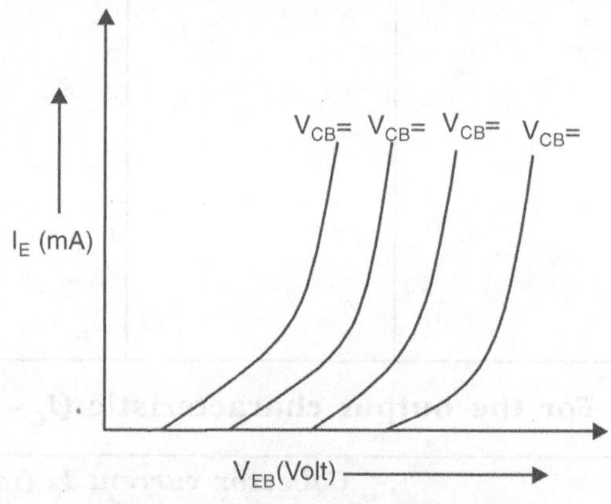

Fig. 33.4. *Input characteristics.*

In input characteristics, the curves obtained in this graph are similar to the forward bias characteristics curve of a *p-n* junction diode. This is because in this case the emitter-base region is forward biased. So I_E increases exponentially with input forward biased voltage V_{EB} and it increases slightly with output reverse biased voltage V_{CB}. It is seen from these curves that the emitter current I_E increases rapidly with small increments in the emitter-base voltage V_{EB}. This means that the input resistance to a small signal voltage applied on the emitter is very low. The curves also suggest that the emitter current is almost independent of the collector to base voltage V_{CB}.

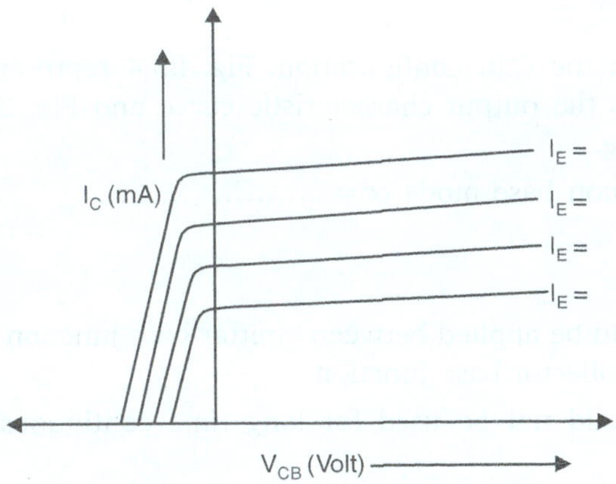

Fig. 33.5. *Output characteristics.*

It is seen from these curves that the entire variation in the collector current I_C takes place at very low value of collector-base (output reverse bias) voltage V_{CB}. Hence further increase in collector-base voltage V_{CB} does not produce appreciable increase in collector current I_C. This current is always a little less than the corresponding emitter current I_E due to the small percentage of charge carriers lost in the thin base due to recombination of holes with electrons.

I_C still flows when V_{CB} is reduced to zero, because I_C is due to the flow of minority charge carriers. I_C becomes zero only after applying forward bias voltage in the collector-base region. In this case I_C does not depend on the value of I_E.

The curves also show that a very large change in the collector-base voltage V_{CB} produces only a small change in the collector current. This means that the output resistance of the transistor is very high. This, in fact causes voltage and power amplification in transistor.

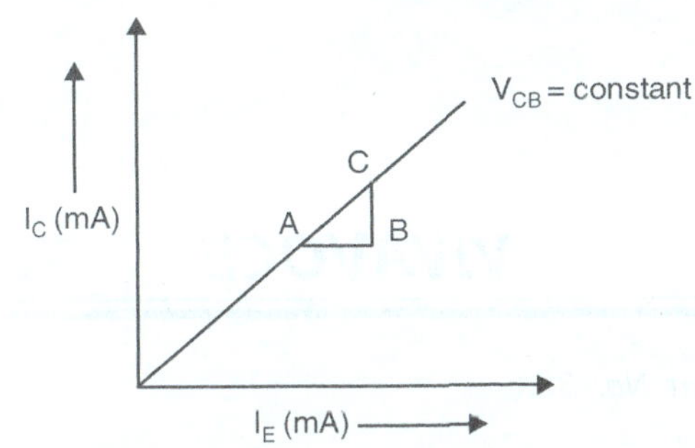

Fig. 33.6. *Current transfer characteristics.*

CALCULATION

Current gain in common base mode of the given transistor can be calculated by the slope of the graph (current transfer characteristic graph).

Current Gain :

$$\alpha = \left(\frac{\Delta I_C}{\Delta I_E}\right)_{V_{CB}} = \textbf{Slope of the } \mathbf{I_C - I_E} \textbf{ curve} = \frac{CB}{AB} = \ldots\ldots\ldots\ldots\ldots$$

RESULT

For the given transistor, in the C.B. configuration. Fig. 33.4 represents the input characteristic curve; Fig. 33.5 represents the output characteristic curve and Fig. 33.6 represents the current transfer characteristic curve.

Current gain in common base mode α = ...

PRECAUTIONS

1. Forward bias should be applied between emitter-base junction and reverse bias should be applied between collector-base junction.
2. The transistor should not be used for long time continuously, it may damage due to heating.

CONCLUSION

VIVAVOCE

Please refer to Experiment No. 32.

❑❑❑

EXPERIMENT—34

AIM

To verify various laws of Boolean algebra and DeMorgan's theorems by using combinational Logic gates.

APPARATUS REQUIRED

Logic gate trainer board, input supply, output device, connecting wires.

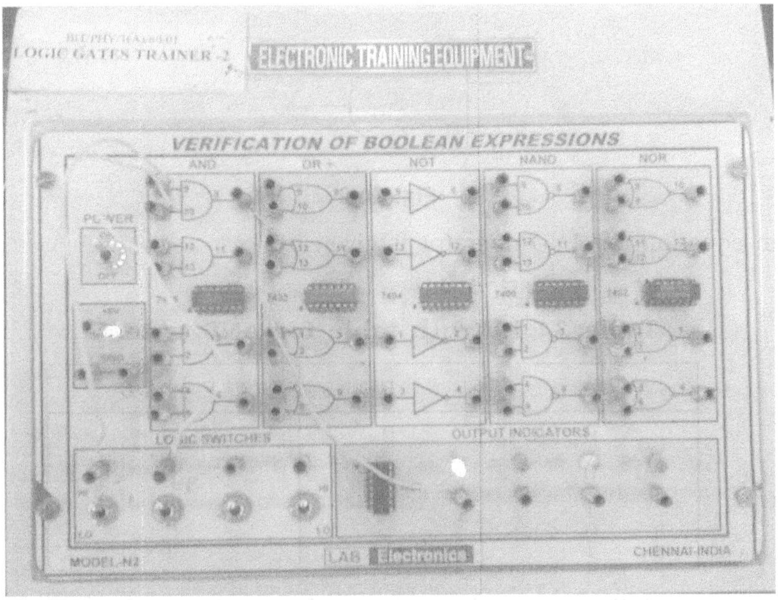

Fig. 34.1. *Logic gate trainer board.*

DESCRIPTION OF THE APPARATUS

Fig. 34.1 shows the experimental kit. It consists of different IC's for performing different logical operations. Switches can vary the inputs. When the switch is at low, the input is considered as 0 and when switch is at high, the input is considered as 1. LED can observe the output. If LED glow it means the output is 1, if it do not glow, the output is 0. The connecting wires are used for connecting the inputs of the gates with the switches.

PRINCIPLE

A digital circuit using transistors and diodes, which makes logical decisions regarding the existence of output depending on the nature of the input, is called logic gate. It has one or more input signals but only one output signal. It performs logical functions using the binary symbols 1 and 0 and a truth table or Boolean algebra describes its performance. There are three types of logic gates, namely (*i*) OR (*ii*) AND (*iii*) NOT operator.

THEORY

Electronic circuits are classified into two parts, namely analog and digital circuits. Analog circuits are those circuits in which voltages and current vary continuously throughout same range. Digital circuits are those circuits in which the output voltage may have only two values either high voltage or low voltage. Digits 1 and 0 represent these two states.

AND Operator

The AND operator is defined for two integer A and B as follows.

"AND operator is such an operator for which output is in logic 1 state, when both inputs A and B are in logic 1 state, otherwise the output is 0." The symbol "." is used to represent AND operation.

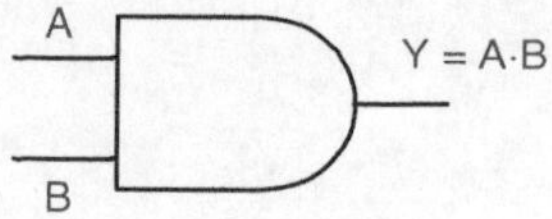

Fig. 34.2

The Boolean equation for AND operator is given by

$$Y = A \cdot B$$

Table 34.1. Truth table for AND gate.

A	*B*	*Y = A·B*
0	0	0
0	1	0
1	0	0
1	1	1

The operation of an AND gate can be understand by a simple electrical circuit Fig. 34.3 in which a bulb is connected to a battery with switch 1 and switch 2 in series. It can be seen that when both the switches (inputs), A and B are in together closed (logic 1 state), the bulb (output) Y will glow (logic 1 state). If one of the switch (input) or both switches (inputs) are open (logic 0 state), the bulb (output) will not glow (logic 0 state).

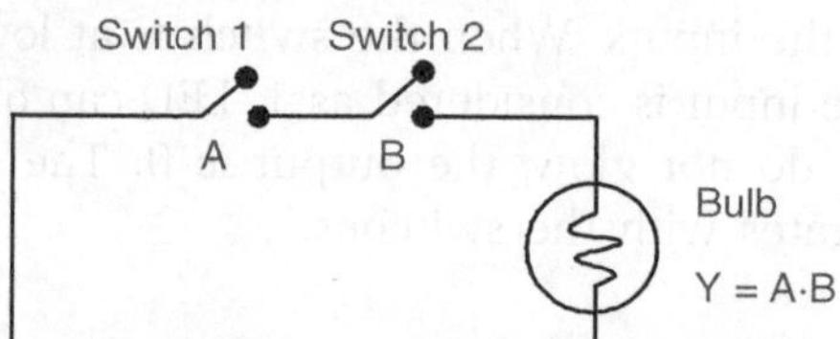

Fig. 34.3. *Electrical circuit for AND gate.*

OR Operator

The OR operator is defined for two integer A and B as follows.

"OR operator is such an operator for which output is in logic 0 state, when both inputs A and B are in logic 0 state, otherwise the output is 1. The symbol "+" is used to represent OR operation.

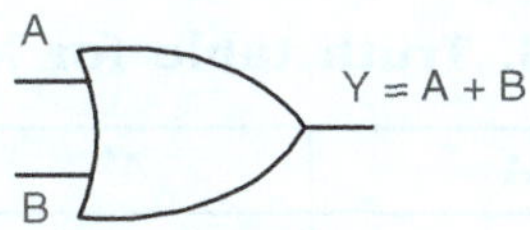

Fig. 34.4

The Boolean equation for OR operator is given by : Y = **A+B**

Table 34.2. Truth table for OR gate.

A	*B*	*Y = A + B*
0	0	0
0	1	1
1	0	1
1	1	1

The operation of OR gate can be understand by a simple electrical circuit Fig. 34.5 in which a bulb is connected to a battery with 2 switches 1 and 2 in parallel. It can be seen that when the switches (inputs), 1 and 2 are in together open (logic 0 state), the bulb (output) Y will not glow (logic 0 state). If one of the switch (input) or both switches (inputs) are closed (logic 1 state), the bulb (output) will glow (logic 1 state).

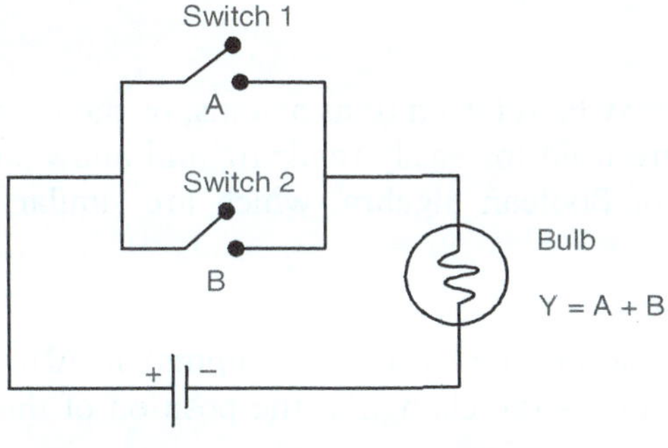

Fig. 34.5. *Electrical circuit for OR gate.*

NOT Operator

It is also called inverter because it only invert the input, that means logic 1 state at the input becomes logic 0 at the output and vice versa. For example, when input A = 0, the output Y = 1 and when A = 1 the output Y = 0.

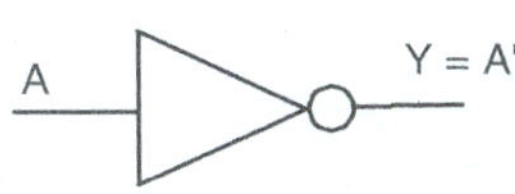

Fig. 34.6

The symbolic representation for NOT operator is given by

$$A = A'$$

Table 34.3. Truth table for NOT gate.

A	*A′*
0	1
1	0

The operation of NOT gate can be understand by an electric circuit as shown in Fig. 34.7 in which a bulb is connected with a photo switch in series. When the bulb is on (logic 1 state), the switch is open (logic 0 state). But when the bulb is off (logic 0 state), the switch is closed (logic 1 state).

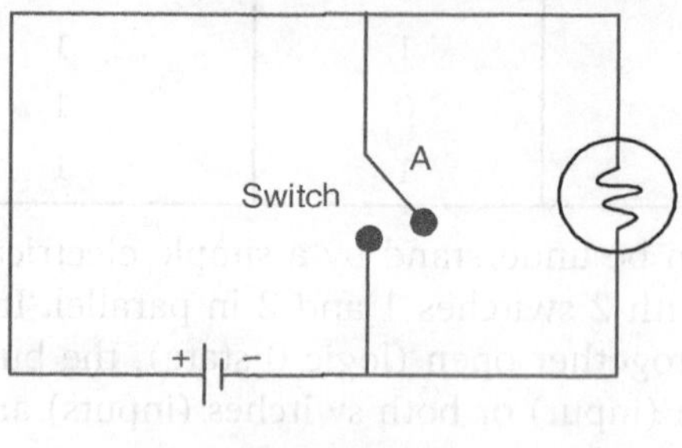

Fig. 34.7. *Electrical circuit for NOT gate.*

Laws of Boolean Algebra

The laws of Boolean algebra may be referred to as axioms, or theorems, which mean that they can be proven to be true. They are used for easily realizing and allow simplification of logic circuits. There are some basic laws of Boolean algebra, which are similar to the operation of normal algebra.

1. **Commutative Law :**

   This law states that the order of variables (inputs) in AND or OR logic gates make no difference. This law allows the change in the position of the logical variables in ANDing and ORing.

   Law-1 : A·B = B·A (For AND gate)

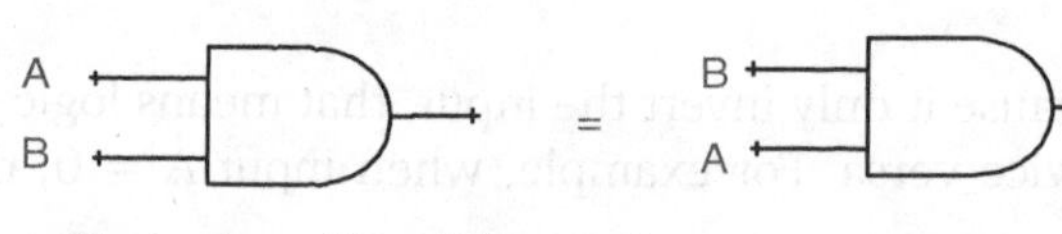

   Law-2 : A + B = B + A (For OR gate)

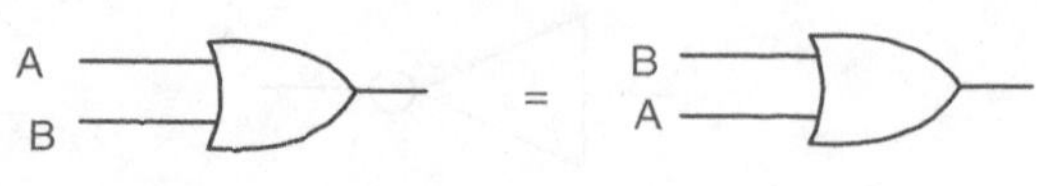

2. **Associative Law :**

   This law allows the grouping of variables. It is similar to the commutative law. The associative laws can be stated as follows :

Law-3 : (A·B)·C = A·(B·C) (For AND gate)

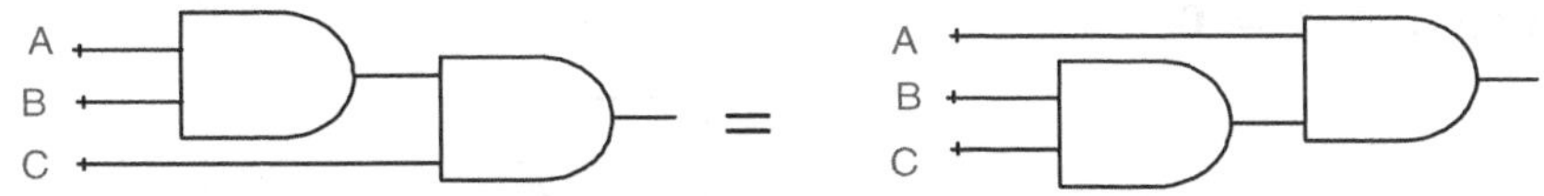

Similarly, for ORing three or more variables, associative law can be written as,

Law-4 : (A + B) + C = A + (B + C) (For OR gate)

3. Distributive Law :

Distributive law shows that factoring or multiplication of different terms in an expression is allowed. According to this law ORing B or C then ANDing with A is equal to ANDing A and B, ANDing A and C then ORing AB or AC. It is similar to normal algebra factoring. Or

Law-5 : A·(B + C) = A·B + A·C

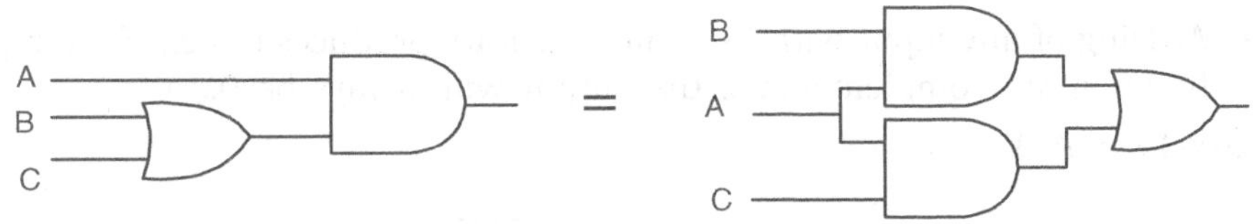

Similarly, ANDing B and C then ORing with A is equal to ORing A or B, ORing A or C then ANDing A + B and A + C. Or

Law-6 : A + (B·C) = (A + B) · (A + C)

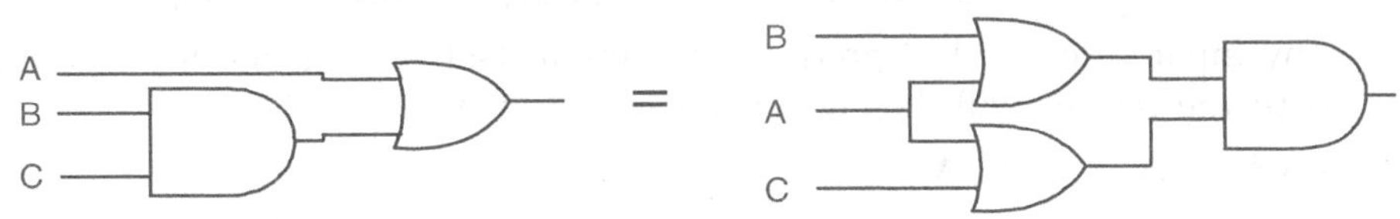

4. Complementary Laws :

This law states that A is not equal to A′ or

Law-7 : A ≠ A′

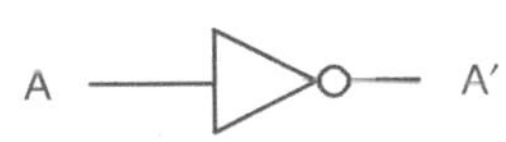

Law-8 : A″ = A

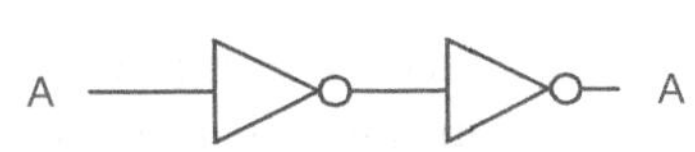

5. AND Laws :

There are following four AND laws which specify the use of ANDing operation itself.

❖ Any input ANDed with 0 produces 0 (zero).

Law-9 : A·0 = 0

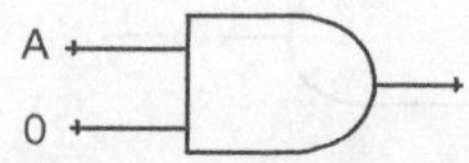

❖ ANDing of any input with 1 produces the input itself, *i.e.,* the output depends on the value of A (0 or 1). It is also known as the law of intersection.

Law-10 : A·1 = A

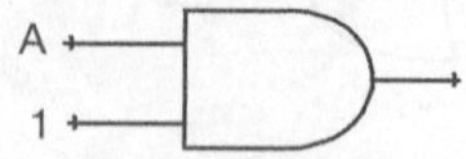

❖ ANDing of equal inputs produces the input itself, *i.e.,* if both inputs to an AND function are same, the output will be the same.

Law-11 : A·A = A

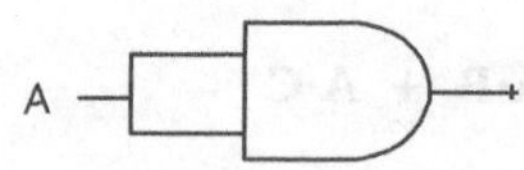

❖ ANDing of any input with its complementary produces 0, *i.e.,* if the inputs to an AND function are complementary, the output will always be 0.

Law-12 : A·A' = 0

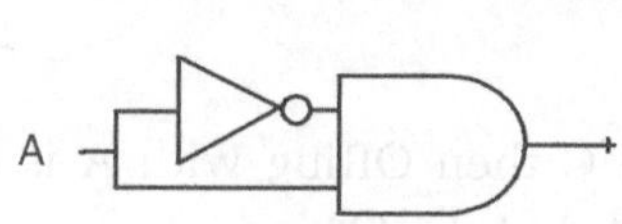

6. OR Laws :

There are following four OR laws which specify the use of ORing operation itself.

❖ Any input ORed with 0 produces the input itself, *i.e.,* if one input to an OR function is 0; the output will depend on the value of A (0 or 1).

Law-13 : A + 0 = A

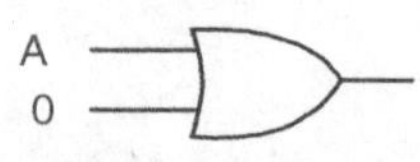

❖ ORing of any input with 1 produces 1, *i.e.,* if one input to an OR function is 1; the output will always be 1.

Law-14 : A + 1 = 1

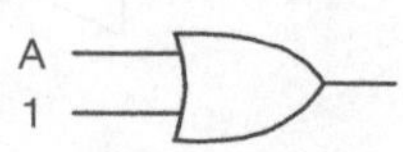

❖ ORing of equal inputs produces the input itself *i.e.,* when both inputs to an OR function are the same, the output will also be the same.

Law-15 : A + A = A

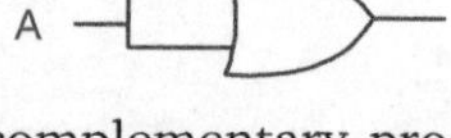

❖ ORing of any input with its complementary produces 1, *i.e.,* if the two inputs to an OR function are complementary, the output will always be 1.

Law-16 : A + A' = 1

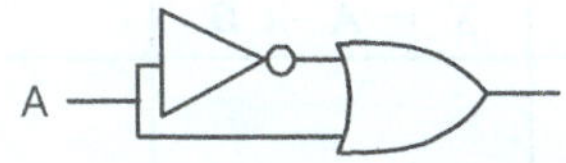

7. **DeMorgan's Theorems.** It states in general the component of any expression can be obtained by replacing each variable and element with its complement and changing OR operators with AND operators and AND operators to OR operators. These theorems can be expressed as follows :
   - ❖ This theorem states negated output AND function is equal to negated input OR function. Or

Law-17 : (A·B)′ = A′ + B′

- ❖ This theorem states that negated output OR function is equal to negated input AND function. Or

Law-18 : (A + B)′ = A′·B′

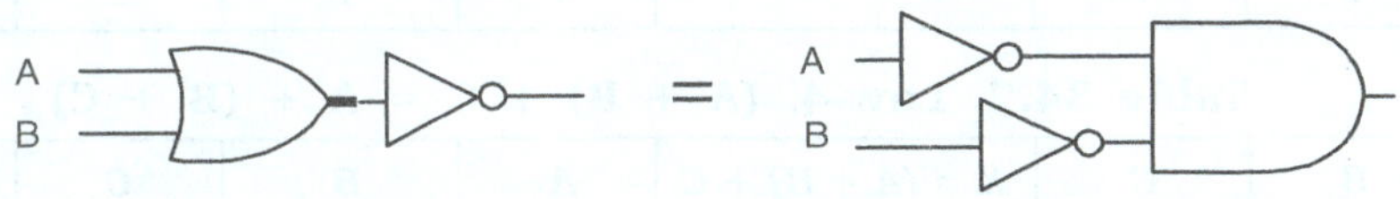

PROCEDURE

1. Addition operation in Boolean algebra is equivalent to the ORing in the logic circuits and multiplication in the Boolean is same as ANDing and inversion is equivalant to NOT operation in the Logic Circuit.
2. High state in circuit is same as 1 logic level and low state is same as 0 logic level. Similarly glowing LED shows the 1 logic level in the output and switched off LED shows 0 logic level in the output.
3. To verify any law, keeping the main switch off the inputs are provided to the gates from input terminals and the output of the logic circuit is connected to any of the detector LED.
4. Now switching on the main switch keeping all the inputs at low position the output is found first and then inputs are changed according to the truth table for that specific rule and the outputs are noted down in the table.

OBSERVATIONS

Table 34.4. Law-1. A·B=B·A.

A	*B*	*X = A·B*	*B*	*A*	*Y= B·A*
0	0		0	0	
1	0		1	0	
0	1		0	1	
1	1		1	1	

Table 34.5. Law-2. A + B=B + A

A	*B*	*X = A + B*	*B*	*A*	*Y = B + A*
0	0		0	0	
1	0		1	0	
0	1		0	1	
1	1		1	1	

Table 34.6. Law-3. (A·B) · C = A · (B·C)

A	*B*	*C*	*X = (A·B)·C*	*A*	*B*	*C*	*Y = A·(B·C)*
0	0	0		0	0	0	
1	0	0		1	0	0	
0	1	0		0	1	0	
0	0	1		0	0	1	
1	1	0		1	1	0	
1	0	1		1	0	1	
0	1	1		0	1	1	
1	1	1		1	1	1	

Table 34.7. Law-4. (A + B) + C = A + (B + C)

A	*B*	*C*	*X = (A + B) + C*	*A*	*B*	*C*	*Y = A + (B + C)*
0	0	0		0	0	0	
1	0	0		1	0	0	
0	1	0		0	1	0	
0	0	1		0	0	1	
1	1	0		1	1	0	
1	0	1		1	0	1	
0	1	1		0	1	1	
1	1	1		1	1	1	

Table 34.8. Law-5. A · (B + C)=A·B + A·C

A	*B*	*C*	*X = A·(B + C)*	*A*	*B*	*C*	*Y = A·B + A·C*
0	0	0		0	0	0	
1	0	0		1	0	0	
0	1	0		0	1	0	
0	0	1		0	0	1	
1	1	0		1	1	0	
1	0	1		1	0	1	
0	1	1		0	1	1	
1	1	1		1	1	1	

Table 34.9. Law-6. A + (B·C)=(A + B) · (A + C)

A	*B*	*C*	*X=A + (B·C)*	*A*	*B*	*C*	*Y = (A + B)·(A + C)*
0	0	0		0	0	0	
1	0	0		1	0	0	
0	1	0		0	1	0	
0	0	1		0	0	1	
1	1	0		1	1	0	
1	0	1		1	0	1	
0	1	1		0	1	1	
1	1	1		1	1	1	

Table 34.10. Law-7. A # A′

A	*Y = A′*
0	
1	

Table 34.11. Law-8. A = A″

A	*Y = A″*
0	
1	

Table 34.12. Law-9. A·0 = 0

A	*Y = A·0*
0	
1	

Table 34.13. Law-10. A·1 = A

A	*Y = A·1*
0	
1	

Table 34.14. Law-11. A·A = A

A	*A*	*Y=A·A*
0	0	
1	1	

Table 34.15. Law-12. A·A′ = 0

A	*A′*	*Y=A·A′*
0	1	
1	0	

Table 34.16. Law-13. A + 0 = A

A	*Y = A+0*
0	
1	

Table 34.17. Law-14. A + 1 = 1

A	*Y = A+1*
0	
1	

Table 34.18. Law-15. A + A = A

A	*A*	*Y = A + A*
0	0	
1	1	

Table 34.19. Law-16. A + A′ = 1

A	*A′*	*Y = A + A′*
0	1	
1	0	

Table 34.20. Law-17. (A·B)′ = A′ + B′

A	*A*	*X = (A·B)′*
0	0	
1	0	
0	1	
1	1	

&

A	*B*	*Y = A′ + B′*
0	0	
1	0	
0	1	
1	1	

Table 34.21. Law-18. (A + B)′ = A′·B′

A	B	X = (A + B)′
0	0	
1	0	
0	1	
1	1	

&

A	B	Y = A′·B′
0	0	
1	0	
0	1	
1	1	

RESULT

The Truth Tables (34.4 to 34.21) for the laws starting from 1 to 18 show the verification of laws.

PRECAUTIONS

1. Input and output connections should be proper.
2. The gate IC's should not be touched by finger.

CONCLUSION

VIVAVOCE

1. What is positive and negative logic ?
2. Explain binary number system.
3. What are the uses of logic gates ?
4. What is truth table ? What does it indicate ?
5. How NAND and NOR gates can be used as universal building block ?
6. What do you mean by gates ?
7. What does digital circuit mean ?
8. What is the difference between digital and analog circuit ?
9. Which are the universal gates and why ?
10. What do you mean by Boolean algebra ?
11. Write and explain Boolean expression for AND, OR and NOT gates.
12. What are DeMorgan's theorems ?
13. Explain how NOT, AND and OR operations can be performed by using NAND or NOR gates.
14. Explain commutative, associative and distributive law of Boolean algebra.
15. Draw the electronic equivalent circuits of each gate.

❑❑❑

APPENDICES

SCHUSTER'S METHOD (focusing the telescope and collimator for parallel rays) :

The slit of collimator is illuminated with sodium light and the prism is placed on the prism table such that its one refracting face AB is towards the collimator and other refracting face AC is towards the telescope. The refracted image of the slit is obtained in the telescope. Then the telescope is turned to bring the spectrum in the field of view of telescope. Now the prism table is turned towards collimator. The spectrum also begins to move in the same direction. The telescope is also moved to keep the spectrum in the same direction. The telescope is also moved to keep the spectrum in the field of view. On turning the prism table towards the collimator, a stage is reached when the spectrum begins to move in opposite direction (*i.e.*, away from collimator). This position of prism is called the position of minimum deviation. The prism table is clamped in this position.

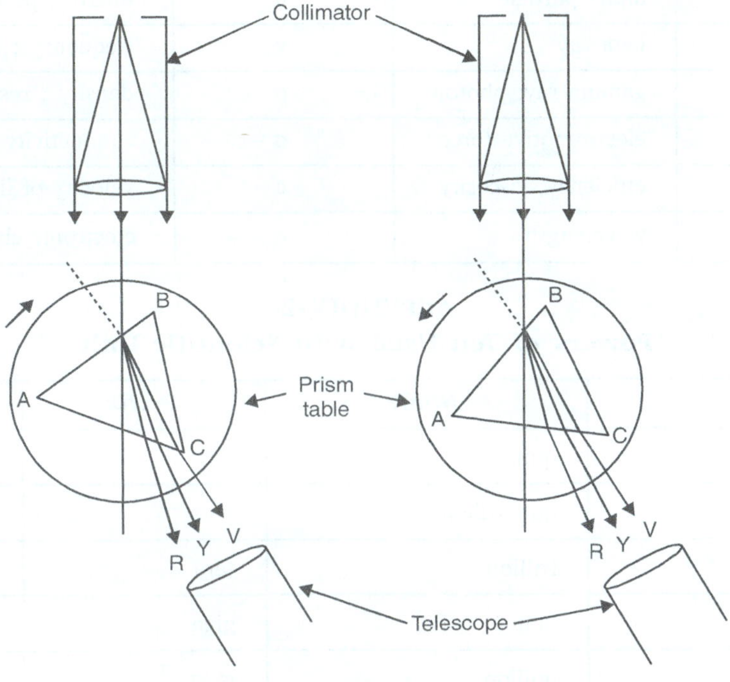

Fig. 1

The telescope is then clamped in a position of slightly greater deviation than the minimum for yellow ray. The prism table is rotated to move the refracting edge of the prism towards the telescope and the telescope is adjusted with the help of its rack and pinion arrangement such that spectrum is distinctly seen. When we move the refracting edge towards the telescope, the angle of incidence becomes greater than the angle of minimum deviation. Therefore emergent beam entering telescope is more parallel.

Now the refracting edge of the prism is slightly turned opposite from the position of minimum deviation (*i.e.*, towards the collimator) and the collimator is adjusted with the help of its rack and pinion arrangement such that spectrum is distinctly seen. When the refracting edge is moved towards the collimator the angle of incidence becomes less than the angle of minimum deviation, with the result that emergent beam is less parallel. Therefore to attain the same parallelism as before, the collimator is adjusted to give the sharp image. The prism is then turned back to the first position and the telescope is adjusted again to obtain the sharp image. Thus, if these two processes are repeated the telescope receives more and more parallel beam and collimator gives more and more parallel beam.

This focuses the telescope and collimator for parallel rays. We take readings on two verniers in order to eliminate the error arising out of the fact that the axis of rotation of the prism table or the telescope may not coincide with the center of the circular main scale. The two readings should differ by 180°.

APPENDIX–1

Physics Symbols and Standard Values of Some Physical Parameters

(source-http://physics.nist.gov/cuu)

Symbol	*Meaning*	*Symbol*	*Meaning*
α	alpha particle	μ	micro- ; permeability
β	beta ray	ν	frequency ; neutrino
γ	gamma ray; photon	ρ	density ; resistivity
ε	electromotive force	σ	conductivity
η	efficiency; viscosity	c	velocity of light
λ	wavelength	e	electronic charge

APPENDIX–2

Powers of Ten Used with Scientific Unit

Factor	*Name*	*Prefix*	*Symbol*
10^{18}	quintillion	exa	E
10^{15}	quadrillion	peta	P
10^{12}	trillion	tera	T
10^{9}	billion	giga	G
10^{6}	million	mega	M
10^{3}	thousand	kilo	K
10^{2}	hundred	hecto	h
10^{1}	ten	deca	da
10^{-1}	one tenth	deci	d
10^{-2}	one hundredth	centi	c
10^{-3}	one thousandth	milli-	m

10^{-6}	one millionth	micro-	μ
10^{-9}	one billionth	nano-	n
10^{-12}	one trillionth	pico-	p
10^{-15}	one quadrillionth	femto-	f
10^{-18}	one quintillionth	atto-	a

APPENDIX–3
Densities (at 20°C)

Substance	*Density* $\times 10^3$ *kg/m*^{3}	*Substance*	*Density* $\times 10^3$ *kg/m*^{3}
Liquids		**Solids**	
Water	1	Brass	8.4
Kerosene	0.83	Copper	8.89
Castor oil	0.97	Aluminium	2.7
Glycerine	1.26	Gold	19.3
Benzene	0.88	Silver	10.5
Ether	0.74	Platinum	21.45
Alcohol	0.8	Antimony	6.6
Mercury	13.6	Chromium	6.92
Aniline	1.02	Iron	7.88

APPENDIX–4
Units of Measurement

Metric unit	*Equivalent*
Length	
1 cm	= 10 mm = 0.3937 in.
1 m	= 100 cm = 1000 mm = 10^6 μm = 10^9 nm
1 m	= 3.281 ft = 39.37 in.
1 km	= 1000 m = 0.6214 mi
1 in.	= 2.540 cm
1 ft.	= 30.48 cm
1 yd	= 91.44 cm

1 mi	= 5280 ft = 1.609 km
1 Å	$= 10^{-10}$ m $= 10^{-8}$ cm $= 10^{-1}$ nm
1 light year	$= 9.461 \times 10^{15}$ m
Mass	
1 kg	1,000 g = 0.0685 slug
1 tone	1,000 kg
Area	
1 cm^2	100 mm^2
1 m^2	10,000 cm^2
1 hectare	10,000 m^2
Volume	
1 cubic centimeter (cc)	1 milliliter
1 liter	1,000 milliliters
1 cubic meter	1,000 liters
1 ft^3	28.32 liters
1 gallon	3.788 liters
Capacity (liquid and dry measures)	
1centiliter	10 ml
1 deciliter	10 cl
1 liter	10 dl
1 decaliter	10 liters
1 hectoliter	10 decaliters
1 kiloliter	10 hectoliters
Time	
1 min	= 60 s
1 h	= 3600 s
1 d	= 86,400 s
1 y	$= 3.156 \times 10^7$ s
Angle	
1 rad	$57.30° = 180°/\pi$
1°	0.01745 rad $= \pi/180$ rad
1 revolution	$= 360° = 2\pi$ rad
1 rpm	= 0.1047 rad/s
Force	
1 N	$= 10^5$ dyn = 0.2248 lb
1 lb	$= 4.448$ N $= 4.448 \times 10^5$ dyn

Pressure	
1 Pa	1 N/m^2 = 1.450 × 10^{-4} $lb/in.^2$ = 0.209 lb/ft^2
1 bar	10^5 Pa
1 atm	1.013 × 10^5 Pa
1 mm Hg	1 torr = 133.3 Pa
Energy	
1 J	10^7 ergs = 0.239 cal
1 cal	4.186 J
1 eV	1.602 × 10^{-19} J
1 kWh	3.600 × 10^6 J
Mass - Energy equivalence	
1 kg	8.988 × 10^{16} J
1 u	931.5 MeV
1 eV	1.074 × 10^{-9} u
Power	
1 W	1 J/s
1 hp	746 W=550 ft.lb/s

APPENDIX–5
Geometry

Area	*Volume*
Circle–π (radius)^2	Cylinder–π (radius)^2 × height
Triangle– ½ × base × height	Cone– 1/3 × π(radius)^2 × side
Rectangle– side × side	Rectangular block– side *a* × side *b* × side *c*
Cylinder– 2π × (surface radius) × height	Cone– π × (surface radius) × side
Rectangular block– 2 × (side *a* × side *b* + side *b* × side *c* + side *c* × side *a*)	

APPENDIX–6
Temperature Scales

To convert from Celsius (C) to Fahrenheit (F) : F = (C × 9/5) + 32

To convert from Fahrenheit (F) to Celsius (C) : C = (F–32) × 5/9

To convert from Celsius to Kelvin (K) : K = C + 273

To convert from Kelvin to Celsius : C = K–273

APPENDIX–7
Elastic Constants

Material	*Young's modulus Y 10^{11} N/m^2*	*Shear modulus η 10^{11} N/m^2*	*Bulk modulus B 10^{11} N/m^2*	*Poisson's ratio σ*
Aluminum	0.70	0.30	0.70	0.16
Brass	0.91	0.36	0.61	0.26
Copper	1.1	0.42	1.4	0.32
Iron	1.9	0.70	1.0	0.27
Steel	2.0	0.84	1.6	0.19
Tungsten	3.6	1.5	2.0	0.20

APPENDIX–8
Surface Tension of Liquids at 20°C

Liquid	*Surface tension N/m*	*Liquid*	*Surface tension N/m*
Mercury	0.465	Water	0.075
Soap solution	0.030	Glycerine	0.063
Carbon tetrachloride	0.027	Ethyl alcohol	0.022

APPENDIX–9
Contact Angles

Substance	*Contact angle*	*Substance*	*Contact angle*
Water with glass	0	Mercury with glass	140°
Water with paraffin	107°	Methylene iodide with glass	29°

APPENDIX–10
Coefficient of Viscosity

Temperature °C	*Viscosity of castor oil, poise*	*Viscosity of water, centipoises*	*Viscosity of air, micropoise*
0	53	1.792	171
20	9.86	1.005	181
40	2.31	0.656	190
60	0.80	0.469	200
80	0.30	0.357	209
100	0.17	0.284	218

APPENDIX–11
Wavelength Range of Colors

Color	*Wavelength*	*Color*	*Wavelength*
Red	620–780 nm	Green	500–570 nm
Orange	590–620 nm	Blue	450–500 nm
Yellow	570–590 nm	Violet	380–450 nm

APPENDIX–12
Index of Refraction for Various Substances Measured with Light of Vacuum Wavelength λ_0 = 589 nm

Substance	*Value*	*Substance*	*Value*
Solids at 20°C		**Liquids at 20°C**	
Crown glass	1.52	Benzene	1.501
Flint glass	1.66	Ethyl alcohol	1.361
Cubic zirconia	2.21	Glycerine	1.473
Diamond	2.419	Water	1.333
Fluorite(CaF_2)	1.434		
Fused quartz(SiO_2)	1.458	**Gases at 0°C**	And 1 atm
Ice	1.309	Air	1.000293
Polystyrene	1.49	Carbon dioxide	1.0045
Sodium Chloride	1.544		

APPENDIX–13
Numerical Constants
Fundamental Physical Constants

Name	*Symbol*	*Value*
Speed of light	c	2.99792458×10^{8} m/s
Magnitude of charge of electron	e	$1.602176462 \times 10^{-19}$ C
Gravitational constant	G	6.673×10^{-11} Nm^2/kg^2
Planck's constant	h	$6.62606876 \times 10^{-34}$ J/s
Boltzmann Constant	k	$1.3806503 \times 10^{-23}$ J/K
Avogadro's number	N_A	$6.02214199 \times 10^{-23}$ molecules/mol
Gas constant	R	8.314472 J/mol. K
Mass of electron	m_e	$9.10938188 \times 10^{-31}$ kg
Mass of proton	m_p	$1.67262158 \times 10^{-27}$ kg
Mass of neutron	m_n	$1.67492716 \times 10^{-27}$ kg
Permeability of free space	μ_0	$4\pi \times 10^{-7}$ Wb/A .m
Permittivity of free space	$\varepsilon_0 = 1/\mu_0 c^2$	$8.854187817... \times 10^{-12}$ C^2/Nm^2
	$1/4\ \pi\varepsilon_0$	$8.987551787... \times 10^{9}$ Nm^2/C^2

Other Useful Constants

Mechanical equivalent of heat		***4.186 J/cal***
Standard atmospheric pressure	1 atm	1.01325×10^{5} Pa
Absolute zero	0 K	–273.15°C
Electron volt	1 eV	$1.602176462 \times 10^{-19}$ J
Atomic mass unit	1 u	$1.66053873 \times 10^{-27}$ kg
Electron rest energy	$m_e c^2$	0.510998902 MeV
Volume of ideal gas(0°C and 1 atm)		22.413996 liter/mol
Acceleration due to gravity (standard)	g	9.80665 m/s^2

APPENDIX–14
Thermal Conductivities of Some Substances

Substance	*Thermal conductivity (W/m°C)*	*Substance*	*Thermal conductivity (W/m°C)*
Metals (at 25°C)		**Non-metals (approx value)**	
Aluminum	238	Asbestos	0.08
Copper	397	Concrete	0.8
Gold	314	Diamond	2300
Iron	79.5	Glass	0.8
Lead	34.7	Ice	2
Silver	427	Water	0.6
Gases (at 20°C)		Wood	0.08
Air	0.0234		
Helium	0.138		
Hydrogen	0.172		
Nitrogen	0.0234		
Oxygen	0.0238		

APPENDIX–15
Internal Resistance and E.M.F. of Some Cells

Cell	*Anode*	*Cathode*	*E.M.F. (in volt)*	*Internal resistance (in ohm)*
Daniel	Copper	Zinc	1.08	3–4
Leclanche	Carbon	Zinc	1.46	5–6
Lead accumulator	Lead peroxide	Sponzy Lead	2.2	0.1–0.2
Ni-Fe	Nickel hydroxide	Iron	1.35	0.2–0.3

APPENDIX-16

Resistivities and Temperature Coefficients of Resistivity for Various Materials at 20°C

Material	*Resistivity* (Ω.m)	*Temperature coefficient* a [(°C)$^{-1}$]
Silver	1.59×10^{-8}	3.8×10^{-3}
Copper	1.7×10^{-8}	3.9×10^{-3}
Gold	2.44×10^{-8}	3.4×10^{-3}
Aluminium	2.82×10^{-8}	3.9×10^{-3}
Tungsten	5.6×10^{-8}	4.5×10^{-3}
Iron	10×10^{-8}	5×10^{-3}
Platinum	11×10^{-8}	3.92×10^{-3}
Lead	22×10^{-8}	3.9×10^{-3}
Nichrome	1.50×10^{-6}	0.4×10^{-3}
Carbon	3.5×10^{-5}	-0.5×10^{-3}
Germanium	0.46	-48×10^{-3}
Silicon	640	-75×10^{-3}
Glass	$10^{10} - 10^{14}$	
Hard rubber	10^{13}	
Sulphur	10^{15}	
Quartz (fused)	75×10^{16}	

APPENDIX-17

Resistance of Wires of Standard Gauge

S.W.G. No.	*Diameter*	*Resistance per meter length (Ω/m)*			
	(in mm)	*Manganin*	*Copper*	*Nichrome*	*Constantan*
10	3.25	0.051	0.0021	0.085	0.057
12	2.64	0.077	0.0032	0.188	0.086
14	2.03	0.131	0.0054	0.317	0.146
16	1.63	0.204	0.0083	0.497	0.228
18	1.22	0.361	0.0148	0.880	0.405

20	0.914	0.645	0.0260	1.56	0.722
22	1.711	1.08	0.0435	2.59	1.2
24	0.559	1.73	0.070	4.2	1.93
26	0.457	2.58	0.105	6.26	2.89
28	0.374	3.82	0.155	9.27	4.27
30	0.315	5.45	0.222	13.2	6.08
32	0.274	7.18	0.293	17.4	8.02
34	0.234	9.9	0.404	24	11.1
36	0.193	14.5	0.590	35.1	16.2
38	0.152	23.2	0.950	56.5	26
40	0.122	36.3	1.48	88.1	30.6

APPENDIX–18
Wavelength of Spectral Lines in Emission Spectra

Substance	***Color***	***Wavelength (in Å)***	***Substance***	***Color***	***Wavelength (in Å)***
Mercury	Violet	4047	Hydrogen	Red (α)	6563
	Violet	4358		Green (β)	4861
	Blue	4916		Blue (γ)	4340
	Green	4960		Violet (δ)	4102
	Green	5461	Sodium	Yellow (D_1)	5896
	Yellow	5770		Yellow (D_2)	5890
	Yellow	5790	Neon	Yellow	5401
Cadmium	Green	5086		Orange	5852
	Red	6438		Red	6402
Helium	Yellow	5876			

APPENDIX–19
Horizontal Component of Earth's Magnetic Field (H) at Different Places

Place	*H (in gauss)*	*Place*	*H (in gauss)*
Agra	0.3484	Dehradun	0.3315
Aligarh	0.3455	Meerut	0.3389
Allahabad	0.3629	Varanasi	0.3635
Kanpur	0.3628	Indore	0.3716
Lucknow	0.3536	Delhi	0.3453
Mumbai	0.376	Ajmer	0.3461
Gwalior	0.3537	Ujjain	0.3582
Jaipur	0.3471	Ghaziabad	0.3444
Kolikata	0.373	Jabalpur	0.364

APPENDIX–20
Symbols used in Electrical Circuits

Symbol	*Meaning*	*Symbol*	*Meaning*
	Resistance	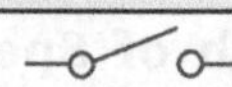	Switch
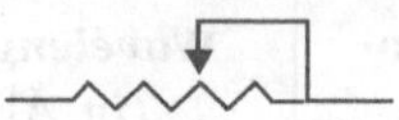	Rheostat	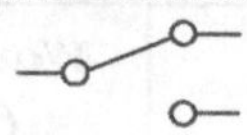	Two Way Switch
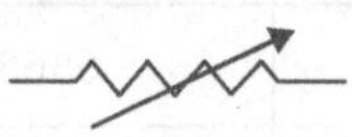		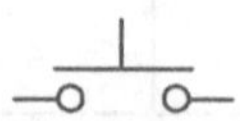	Push Button Switch
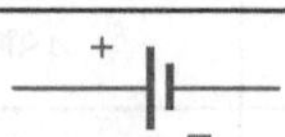	Cell/Battery	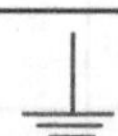	Earth/Ground
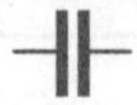	Capacitor/Condenser		Inductor
	AC Source	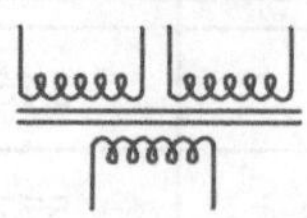	Transformer
	Fuse	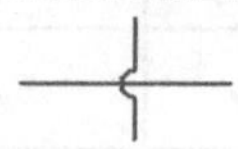	Crossed-wires (Not Connected)
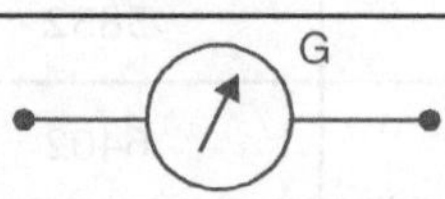	Galvanometer	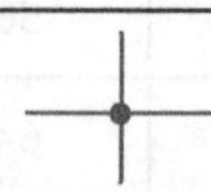	Crossed-wires (Connected)
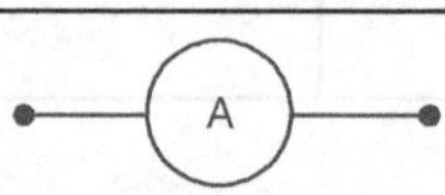	Ammeter	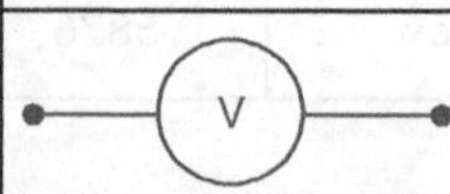	Voltmeter

Index

A

B

C

D

E

F

G

H

I

J

K

L

M

N

O

P

R

S

T

U

V

W

Z

❑❑❑